Definitions, Shapes, and Forces.

Physics: The Art & Math of Building & Moving Things.

Volume 1

by M. P. Schaefer

Definitions, Shapes, & Forces

Physics: The Art & Math of Building and Moving Things.

v.1

ISBN-13:
978-1500130077

ISBN-10:
1500130079

Table of Contents

DEDICATION

I would like to thank my middle school & high school science teachers, who actually knew their subjects and knew how to teach.

I would also like to thank my mother, who taught me to read before I ever got into school by reading to me every night.

1 - Foreword:

Most science books these days are arranged in chronological order describing each discovery, giving credit to the official discoverer, and bringing up formula after formula without much explanation or practical use. It would be better to call them "History of Science" books.

This book, and any others in this series, will be arranged a little differently. I'm starting with easy and simple concepts and building on them, while teaching you the skills to discover & create on your own.

2 - What is Science?:

This might seem too simple, but just to make sure we're on the same page let's take a moment to define it.

The word Science comes from Latin and means knowledge or an area of study. It refers to any organized body of knowledge, particularly those developed using the "scientific method".

The Scientific Method refers to a process of at least 5 steps for figuring out the principles of how things work and confirming the validity of those principles.

1. Research: Literally searching again for any previous documentation about the subject you're studying. In other words reading and studying related information.

2. Hypothesis: Hypo means under. Thesis means an opinion, a proposal, or a theory. In other words an undeveloped theory based on your experiences and research.

3. Experiment: Come up with some method of testing your hypothesis. The type of experiment will depend on the subject you're studying. Surveying people's experiences or opinions is one type of experiment. Immersing an object in water to see if it floats or dissolves or changes temperature (etc.) is another type of experiment. In some cases it helps to have a control group for your experiment. That basically means you have a sample that is left unchanged or treated in a manner that shouldn't change it, so it can be used to compare to your experiment results.

4. Analysis: The word come from the Greek language and means to break something up. In this case it refers to breaking up all the   information from your research and experimentation and organizing it to make it easier to understand.

5. Conclusion: Take all that information and try figure out what it means, and why it works that way, and if your hypothesis was accurate. Sometimes the initial conclusion is that more research or more analysis is necessary.

Other steps may include documentation and publishing your findings.

There are many branches of science. This series covers the science of physics. Physics is derived from a Greek word meaning nature and refers to the study of matter, energy, structure and motion.

Now that we've started speaking the same language, let's start the next chapter with something relatively easy.

3 - How force works:

1. A Force is a push or a pull, or it can also be a rotation or a vibration. It is a source of energy and is considered a "vector quantity", which means it has a magnitude (size or amount) and a direction.

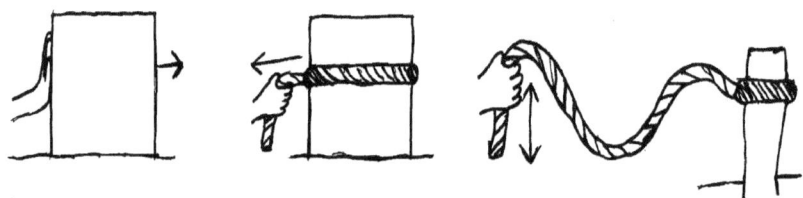

Energy is the ability to do work or produce a change.

2. If the forces effecting an object are balanced, it will not accelerate.

Accelerate means to change velocity. Velocity is the speed something or someone is moving in a particular direction. In other words starting a motion, stopping, speeding up, & slowing down are all accelerations.

3. If the forces on an object are unbalanced, it will accelerate in the same direction of the larger force (also known as the Resultant force).

Principles 2 & 3 make up Newton's 1st Law, the Law of Inertia (Inertia is the tendency of an object to resist changes in motion)

4. The larger the force is the greater the acceleration will be. The larger the mass of the object the smaller the acceleration will be. This relationship can be represented by the equation: acceleration equals force divided by mass ($a=F/m$), or the more common: force equals mass times acceleration ($F=ma$).

Principles 3 & 4 make up Newtons second law.

5. Energy can neither be created nor destroyed, but can be transformed from one kind to another. That's just another way of saying things don't just move/change on their own, there's always a cause.

Principle 5 is called the Law of Conservation of Energy.

6. Matter can neither be created nor destroyed in ordinary reactions, but can be transformed from one form or type to another. Which another way of saying you can't get something from nothing, there must be a source and things don't just disappear, it had to go somewhere.

Principle 6 is called the Law of Conservation of Mass.

7. Mass refers to the amount of matter in an object as measured by it's inertia. The standardized unit of mass for the Metric System is the kilogram. This is the mass of one liter (~33.8 fluid ounces) of water. The Old English unit for mass is the Slug. A slug has enough mass that it will be accelerated at a rate of 1 foot/second for each second a one pound force is applied and is equal to 14.6 kilograms. A one slug object weighs 32.17 pounds.

8. In the Old English System that is still used in America, force is usually measured in pounds. In the Systems International or Metric system it's measured in Newtons. One Newton is the force that will accelerate 1 Liter of water, or anything else with a kilogram of mass, by 1 meter per second for each second it's applied or $1m/sec.^2$. 1 pound = 4.45 Newtons.

9. For every force applied to an object, there's an equal and opposite force effecting another object. If you swing a hammer to hit a nail, a counter pressure is effecting the hand that's holding the hammer, and as the nail is struck, the shock affects the hammer; it's handle; and your hand. Another way to say this is that there are consequences for every action.

Principle 9 is Newton's Third Law, also known as the Law of Action & Reaction.

10. The masses of two objects will attract each other with forces of equal magnitude. An example of this would be the Earth drawing you downward with a force of 100+ pounds. At the same time you are attracting the Earth with the same force.

11. Earth's gravitational force or downward acceleration rate is 9.8 meters [~32.174 feet or ~386.09 inches] per second for each second of falling, or 9.8 m/sec^2. As such the weight of a one   kilogram object would be 9.8 Newtons or 2.2 pounds.

12. Friction is the resistance caused by moving one object or surface while it's in contact with another. All surfaces have Roughness, or minute projections that catch on other surfaces. The greater the connecting surface area, the greater the resistance will be. This is greater when the surfaces are at rest or static [not moving] relative to each other. The coefficient [standardized number] for kinetic[moving] friction equals that friction force(f) divided by the normal force(F_n) being applied between the two [usually the weight]. For static friction's coefficient, The friction force equals the critical force to get the surfaces sliding.

4 - Structure & Shape:

The structure and shape of an object; tool; or building, affects how easily it is effected by different forces and how easy it is to do work.

Looking around you through out the day there are various shapes, from the simple to the complex, both 2 dimensional and 3 dimensional.

Hm. The word dimension has been a cause of confusion for many people over the years so let's take a moment to define it. A Dimension of an object or a space is the minimum amount of coordinates/numbers/ measurements needed to specify a location on or in it.

A straight line has 1 dimension, length. Squares; circles; triangles, etcetera, have 2 dimensions, length & width. Spheres; cubes; cones, etcetera, have 3 dimensions, length; width; & height or depth. A moving object or changing situation has 4 dimensions, with the fourth dimension being time.

Once you go past having a one dimensional straight line, each extra dimension gives you a great deal of information to work with, especially for those that didn't do as well in geometry, so let's take them one at a time.

2D Shapes:

Squares & Rectangles:

The easiest shape to work with in terms of calculations is a square.

A square has 4 corners and four sides and the opposing sides are parallel, just like a parallelogram.(parallel is from a Greek word for "beside"; meaning going in the same direction without getting closer or farther, & -o-gram meaning shape of...)

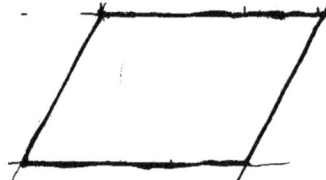

All sides of a square are the same length, just like an equilaterogram/rhombus. (equi-=equal, latero-=side)

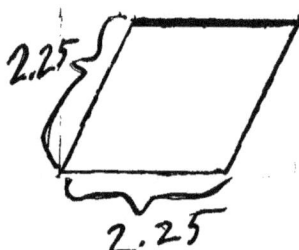

All the corners of a square are at right angles or 90 degrees, just like a rectangle. (rect- is from Latin meaning upright or straight)

To accurately draw squares or rectangles, all you need is a ruler; yardstick; or other straight-edge with measurements on it, and a pencil or a pen.

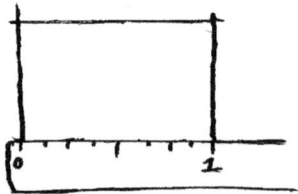

Now there are a couple of important calculations for all 2 dimensional shapes. One is the perimeter or the distance around the outside edge. If you were making a fence or a border around a plot of land or a building, knowing the perimeter size would help you calculate how much material you would need. To calculate you just add the lengths of all the sides.

Math Help 1:

For those of you who are a little slow at math, there's an addition technique that makes it easier to add in your head. It's called Left to Right Addition.

Just like it sounds, you start on the left with the larger numbers. For example if you needed the perimeter of a rectangle that was 225 inches long and 150 inches wide, add the hundred's column first, then the ten's, then the one's.

225	200	620	745
225	200	20	5
150	100	50	0
+150	100	50	0
	600	740	750

With practice you can do it faster than a calculator.

The other calculation is the Surface area, or the total amount of space you have to work with. For squares & rectangles, just multiply the length by the width. In the case of a square it's the same number so multiply that number by itself, which is appropriately called "Squaring", or finding the square of the number. The produced number is represented by square units like square feet, square inches, square meters, & square centimeters.

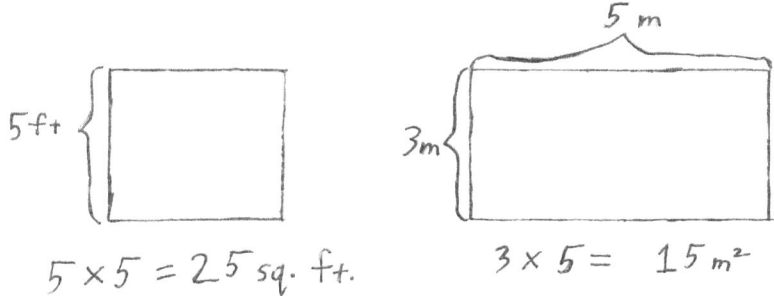

$5 \times 5 = 25$ sq. ft.

$3 \times 5 = 15 \, m^2$

Math Help 2:

There is a trick for mental multiplication called "Cross Multiplication". It can be done from left to right or right to left. There are two points to keep in mind. The first is the place values your working with. If you're multiplying hundreds by hundreds that product is going to be in the tens of thousands' range, and if you're multiplying hundreds by tens it's going to be in the thousands'. The second point is to keep a running total as you run the figures. For example:

125	100	100	300	100	300	20
x324	x300	x20	x20	x 4	x 5	x20
	30,000	+2,000	+6,000	+400	+1,500	+400
		32,000	38,000	38,400	39,900	40,300

20	20	4
x4	x5	x5
+80	+100	+20
40,380	40,480	40,500

Remember, if you see a formula with letters beside each other, the numbers they represent are supposed to be multiplied. If you see such a letter outside some brackets, take care of the equation inside the brackets first, then multiply that answer by the number that letter represents. Sometimes in multiplication problems, you will see an *(asterisk) instead of an x.

A gentleman by the name of Scott Flansburg (his nickname is the Human Calculator) has spent time going around the world, teaching kids to have fun with math shortcuts, and developing a book and an audio/video program to teach them to more people. Most people, especially young kids, can get faster at mental math than a calculator with regular practice.

You can think of surface area as "How many squares of that unit size will fit in that space?".

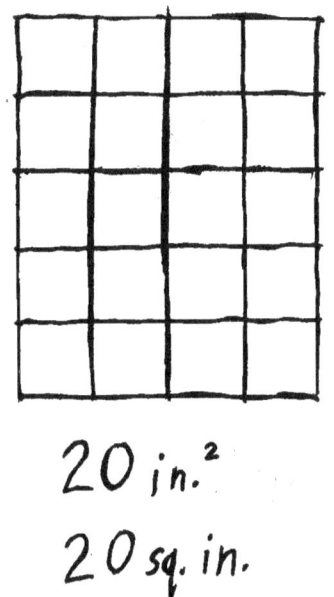

$20 \, in.^2$

$20 \, sq. \, in.$

This can be useful when laying tiles or in calculating pressure. Pressure is the relative distribution of force. If you have a 1 pound force applied across a 1 square inch surface you have a pressure of 1 pound per square inch (psi). Just divide the number of pounds by the area. If you have a 50 pound object and the bottom of it has a surface area of 10 square inches, it's putting 5psi of pressure on the surface it's sitting on (50÷10=5).

Math Help 3:

For those of you who haven't mastered division and fractions yet, here are a few basics.

1. The number being divided (usually the big number) is called the dividend. The number it's divided by is called the divisor. The answer is called the quotient.

2. A fraction represents a division relationship so 50/10 is the same as 50÷10 and 10)50 . The number below or immediately to the right the line is the divisor & is called the denominator (sets the denomination as fifths; sixths; etc.). The number above or immediately to the left of the line is the dividend and is called the numerator.

3. A Proper fraction has a numerator that is smaller than the denominator [e.g.4/5]. An Improper fraction has a numerator that is larger than the denominator [e.g.5/4]. A Mixed fraction combines a fraction and a whole number [e.g.1 ¼].

4. If the dividend and the divisor are both divisible (able to be divided without a remainder) by the same number, you can use that to simplify the problem. For example: 50/10=5/1=5

5. If you divide a fraction by a fraction, you can get the same answer by multiplying it by the reciprocal (reversing the top & bottom, represented by 1/x) of the second fraction. 4/5 ÷ 1/5 = 4/5x5/1 = 4/1 x 1/1 = 4

6. Some numbers have a basic pattern to check if their divisible:

Even numbers are divisible by 2.

If the digits add up to 3; 6; or 9.[: 18(1+8=9)] it's divisible by 3.

If the ten's place is an even number & the one's is 0; 4; or 8, or if the ten's place is odd & the one's is 2 or 6, it's divisible by 4.

If the one's place is 5 or 0, it's divisible by 5.

Divisible by 3 & 2 means it's divisible by 6.

•8: Under 100 or if hundred's place is even, if the ten's place is 0; 4; or 8 the one's is 8 or 0, if 1; 5; or 9 the one's is 6, if 2; 6; or 0 the one's is 4, if 3 or 7 the one's is 2. If the hundred's place is odd add 2 to the ten's on this list.

•9: The digits add up to 9.

•10: The one's place is 0.

•11: For numbers in the hundreds, the ten's place is the sum of the one's and hundred's

•12: divisible by 3 & 4

•15: divisible by 3 & 5

Now on to

Circles:

The second easiest type of shape is a circle.

The most important measurement for a circle is the radius. The Radius is the length a line would have if it were radiating from the center point to the edge of a circle.

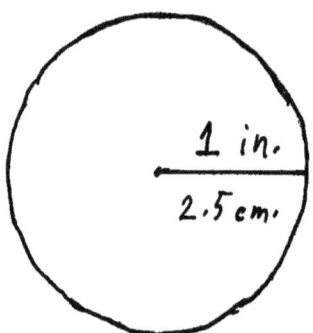

Related to the radius is the diameter (derived from Greek, dia-=across/ through, meter=measure). The Diameter is measured across from edge to edge, through the center, and is always double the radius.

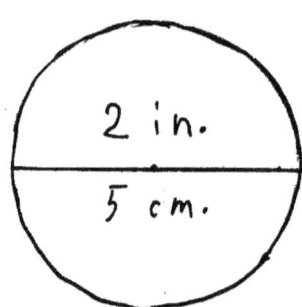

The perimeter of a circle is called the Circumference. You could measure it directly, but it was discovered long ago that the ratio between the circumference and the radius or diameter is always about the same and is represented by the Greek letter π pi(pronounced like pie). The number value of π Pi is 3.14159... and continues on for what seems like forever to the right of the decimal point, but 3.14 is accurate enough for most scientific work. Just multiply the radius by 2 (to get the diameter) and multiply that by π Pi to get the circumference. The abbreviation for this formula is: Circumference=2πR .

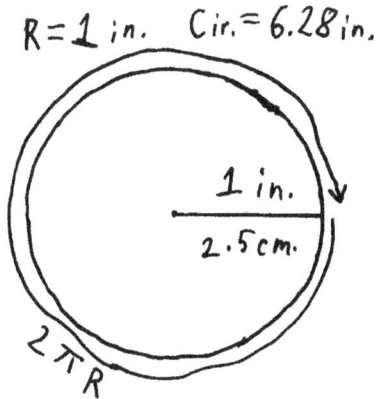

R=1 in.   Cir.=6.28 in.

1 in.

2.5 cm.

2πR

The circumference is important for finding out how fast a vehicle is going. If a wheel has a circumference of 2 meters and it's turning 2 times per second while rolling, the vehicle is travelling at 4 meters per second. To find out how far it would go in 1 minute multiply that times 60 seconds/minute to get 240 meters/minute. To find out how far per hour multiply the minute speed by 60 minutes/hour to get 14400 meters/hour or 14.4 kilometers per hour(kph). To covert that to miles divide by 1.6093 (or round up to 1.61) for about 8.95 miles per hour(mph).

There is a similar calculation to find the surface area of a circle. This time you multiply the radius times itself (squaring for 2 dimensions) and multiply that times π Pi.

The abbreviated formula is: Area=πR².

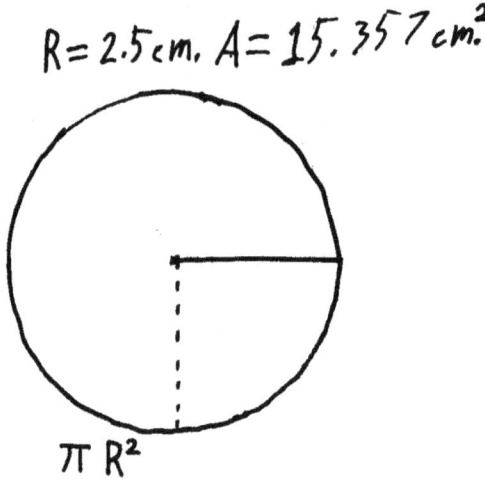

$$R = 2.5 cm, \quad A = 15.357 cm^2$$

$$\pi R^2$$

The area of a circle can be used for figuring how much will fit in a space. It's also important for calculating fluid pressure inside of a pipe.

To accurately draw circles you need a round object to trace or a drawing compass, not to be confused with a magnetic compass for finding direction. A drawing compass usually has a metal point on one arm and a pencil on the other. The two arms are adjustable to set the radius of the circle. For larger circles you can use a string to tie one pencil to another or to a pin or nail at the right distance.

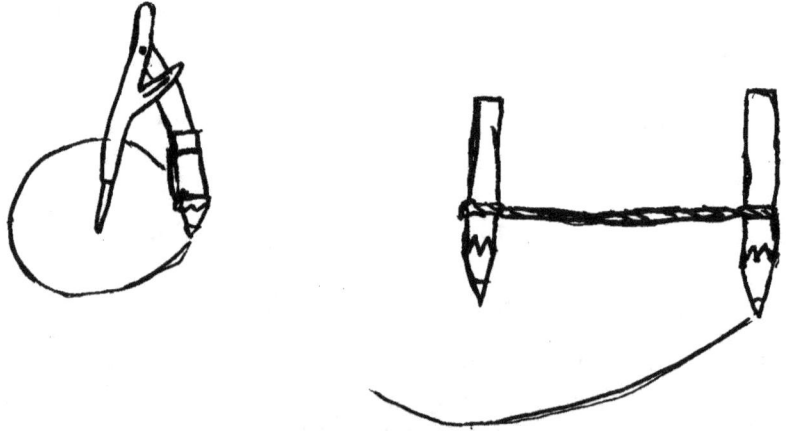

And now....

Triangles:

The next easiest type of shape is a triangle. (Derived from Latin, tri=3, angle= corner)

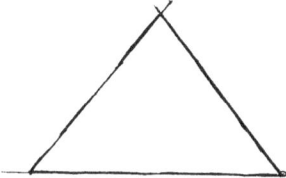

There are a number of different names given to triangles depending on how they're made. An Equilateral triangle has all three sides the same length, and all the angles are 60 degrees(60°).

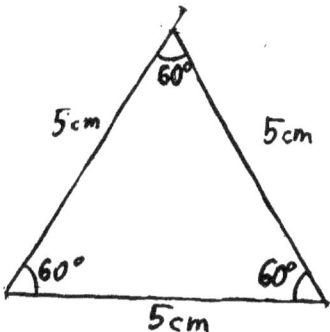

An Isosceles triangle has 2 sides the same length.

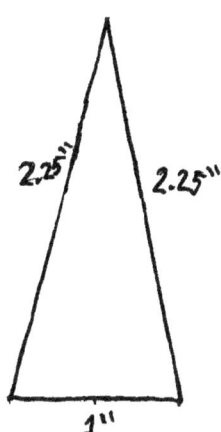

A Scalene triangle has different lengths on all 3 sides.

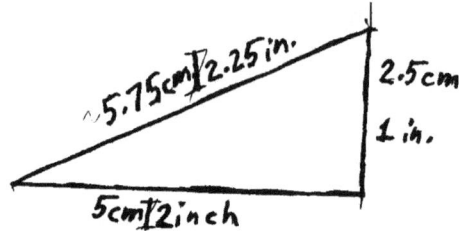

A Right triangle has one angle that is 90°.

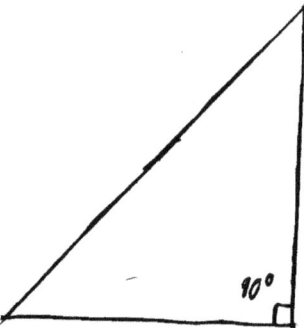

An Acute triangle has all of its corners less than 90°.

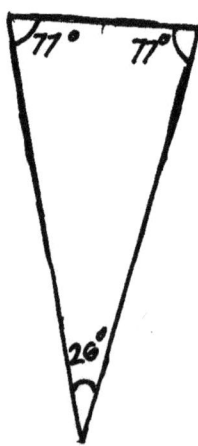

An Obtuse triangle has one corner that's more than 90°.

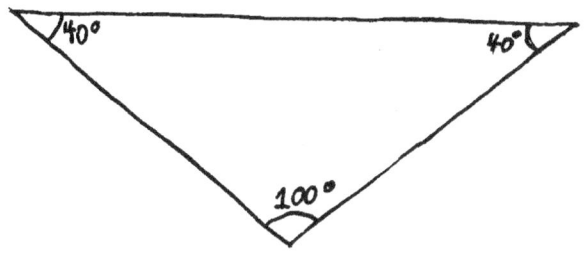

Did you notice? All the angles on a triangle add up to 180°. For four sided shapes, also known as Quadrilaterals (quadri- = 4), all the angles add up to 360°.

In order to draw accurate triangles or measure the angle on an existing triangle, you need a protractor. The most common type these days is a half circle shape that ranges from 0°-180°. It may also measure   centimeters and/or inches.

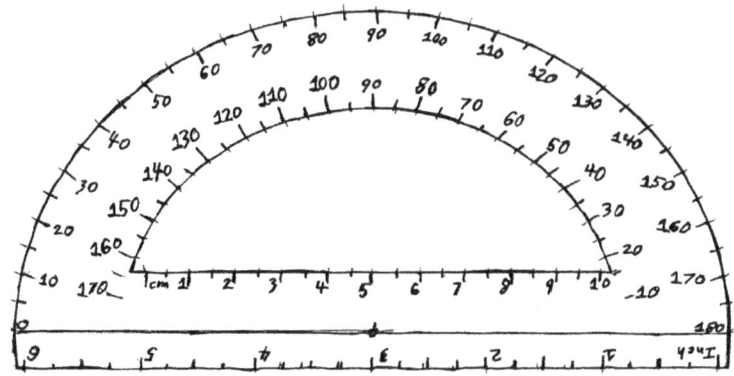

There's another type that's a full circle measuring from 0°-360°.

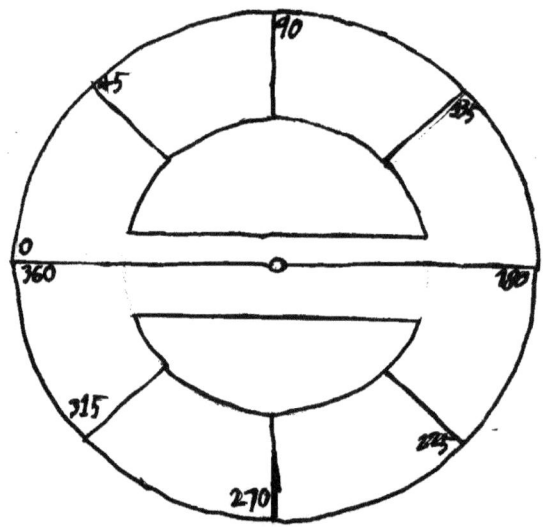

It's no coincidence that the number of degrees in a circle and the number of days in one year on Earth (365) are so close. There were many cultures that used to use a 360 day calender. An extra 30 day month was added every 6 years and either 3 extra days every 12 years or a second extra month added every 120 years.

It isn't used much, but there is another way of measuring angles, called the Radian system. Based on the formula for the circumference of a circle (2πR), there are 6.38 radians in a full circle. So 1 radian (rad) equals about 56.4° (360/6.38).

For the perimeter of a triangle, just add the lengths of the sides. For the area it's a little more complex. Pick one side to use as a base so you can measure the height of the opposite point. One half of the width of the base multiplied by the height gives you the triangle's surface area.

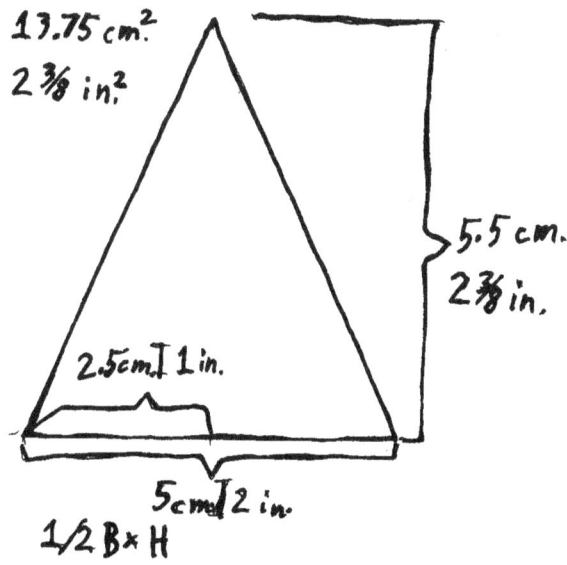

If you draw a square or rectangle around a triangle with the length & width the same size as the height & the base, the part(s) outside your triangle will be the right size & shape to make another triangle with the same measurements.

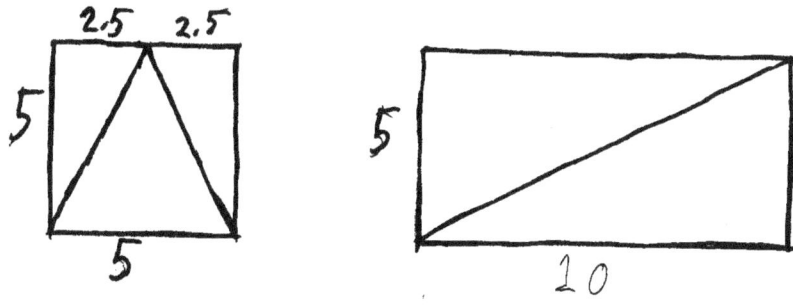

You can also split any 4 sided shape into 2 triangles by drawing a straight line from one corner to the opposite corner.

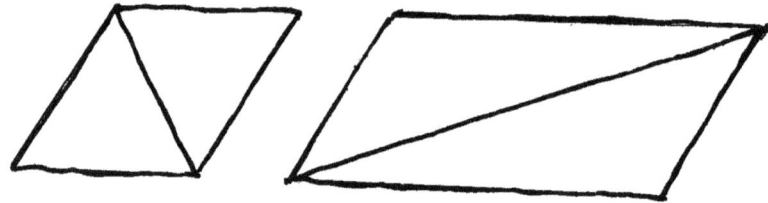

You can make two right triangles out of any other type by drawing a straight line from the base, at 90°, to the opposite point.

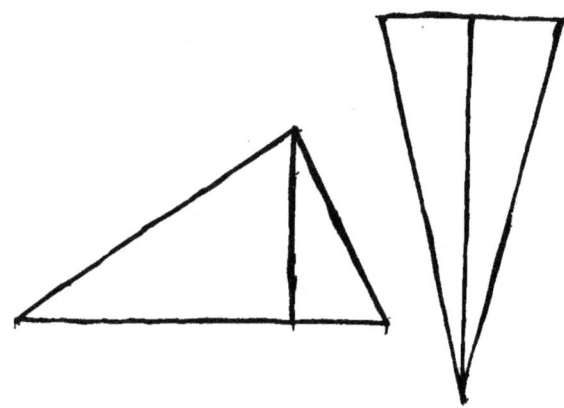

$$A^2 + B^2 = C^2$$

$$6.25\,cm + 25\,cm = 31.25\,cm$$

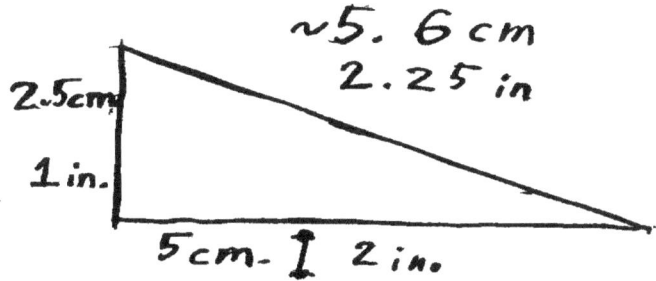

~5. 6 cm
2.25 in

2.5cm

1 in.

5cm. ↕ 2 in.

For right triangles, if you square the length of the short sides and add them together, that sum equals the square of the length of the long side (also known as the Hypotenuse). So if you know the length of 2 sides you can figure out the length of the other side.

Math Help 4:

Even if you don't have a calculator, there is a way of finding square roots. Square root refers to a number that if squared would produce the number you're working with. For example 5 squared equals 25, so 5 is the square root of 25. Before we work on finding the square roots of larger numbers, let's list the squares of 1 through 10.

$1^2=1$, $2^2=4$, $3^2=9$, $4^2=16$, $5^2=25$, $6^2=36$, $7^2=49$, $8^2=64$, $9^2=81$, $10^2=100$

If you try to find the square root of a number that falls in between these with a calculator, you'll get an answer with numbers to the right of a   decimal point. If you multiply non-whole numbers (numbers to the right of a decimal point represent partial units as tenths; hundredths; etcetera) you won't get whole numbers as a product. (0.5x0.5=0.25, 1.5x1.5=2.25) These sub-decimal numbers indicate; how far over the number you were checking was, and will get a product close to it. So the calculations and your measurements won't always match up that accurately, but it will be close enough to work with.

Now to find the square root of a larger number. First, break the number into two digit groups starting from the right side. So if the number you're finding the square root of is 144 (represented as $\sqrt{144}$) , beak it into 1 and 44.

Looking at the left number group, 1, there's only one number that when squared will fit into that so the ten's place of the answer is 1. Looking at the right number group, there are two roots that when squared end in 4, 2 and 8. 8^2 is 64; which is two big, leaving 2 as the likely one's place of the root. So let's check it. 12 x 12 = 144.

Now let's try it on a much larger number, $\sqrt{81,225}$. First separate it into two digit groups. 8, 12, & 25. The largest square that will go into 8 is 4 with a root of 2. So, let's subtract that out.

$$\begin{array}{r} 2 \\ \hline \sqrt{8} \ \ 12 \ \ 25 \\ 4 \\ \hline 4 \end{array}$$

Now bring down the next group so we can figure out the next part of the root.

$$\begin{array}{r} 2 \\ \hline \sqrt{8} \ \ 12 \ \ 25 \\ 4 \\ \hline 4 \ \ 12 \end{array}$$

Since the part of the root we already found affects this part of the square as well, double it & bring it down as a partial divisor & divide it into the first part of the dividend, adding the answer to the root and the divisor as the next digit and ignoring the remainder.

$$\begin{array}{r} 2(6) \\ \hline \sqrt{8} \ \ 12 \ \ 25 \\ 4 \\ \hline \end{array}$$
4(6))412

Now let's multiply the new number (6) by the divisor (46) to see if it's too large or too small.

$$\begin{array}{r} 2(6) \\ \hline \sqrt{8} \ \ 12 \ \ 25 \\ 4 \\ \hline \end{array}$$
46)412
6x46=276

Looks like it'll be too small, so let's try the next larger root.

$$\begin{array}{r} 2(7) \\ \hline \sqrt{8} \ \ 12 \ \ 25 \\ 4 \\ \hline \end{array}$$
47)412
7x47=329

That still seems a little bit small, so let's try the next larger root.

```
        2(8)
    √8  12  25
     4
  48)412
8x48=384
      28
```

Much better. Now let's bring down the next group and double the current square root as a partial divisor to figure out the last part of the root.

```
       28(5)
    √8  12  25
     4
  48)412
     384
 56())28 25
     280
```

Now add that to the divisor as the next digit and multiply it by the divisor to check your work.

```
       28(5)
    √8  12  25
     4
  48)412
     384
 56(5))28 25
      2825
```

Of course since it ended in a 5, as long as it was a square of a whole number, we could have saved ourselves a step and assumed that part of the root was a 5, but it's better to be safe than sorry!

Practical Trigonometry:

Trigonometry (trigo=triangle, nom=number, metry= measuring) uses the predictable angles & measurements of right triangles to calculate distances and forces.

For right triangles, if you know one of the other angles and the length of one of the sides, you can calculate the remaining angle and the other 2 lengths. Because these factors maintain their scale. it's easy to use triangles to identify an object's relative location in a process known as Triangulation.

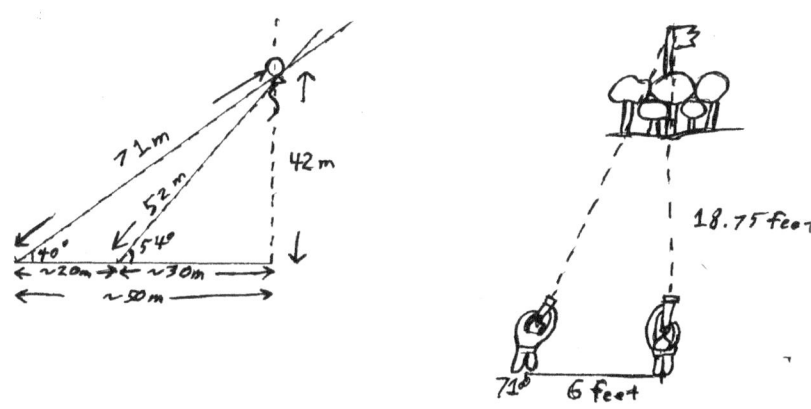

Since the angles on a triangle add up to 180° and one of the angles is 90°, it's very easy once you know one of the other angles to subtract it from 90 to calculate the remaining angle. By taking measurements at different distances and measuring the angles you can draft out a triangle with those angles and assign a convenient scale (i.e.1 inch = 6 feet or 1 centimeter = 10 meters). From this you can measure the unknown distances on your drawing and, using the scale, calculate the actual distance.

It's may be inconvenient to carry a protractor into the field, so in case you need to measure angles from your position relative to objects in your environment, there is a convenient and ancient method. Just stretch out your arm and clench your fist with your wrist at an angle where you can see the back of your knuckles. The width of your fist across the knuckles at that angle takes up about 10° out of your visual field. Need a smaller angle? Stretch out your index or middle finger from that fist. It's about 2° across the tip. You can also use that to find & measure stars & constellations in the night sky.

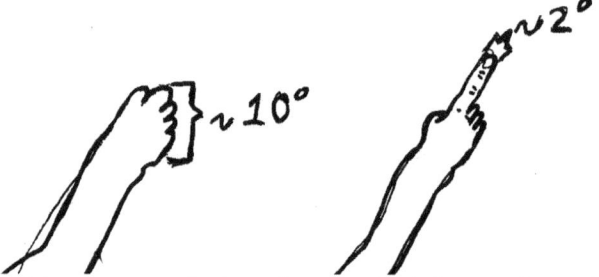

Of course taking paper & drawing materials out into the field and drawing out a scaled right triangle can also be problematic. An easier way would be to use a trigonometric function. The Trigonometric function of an angle is equal to a ratio between the sizes of 2 of the sides of a right triangle that has that as one of it's angles.

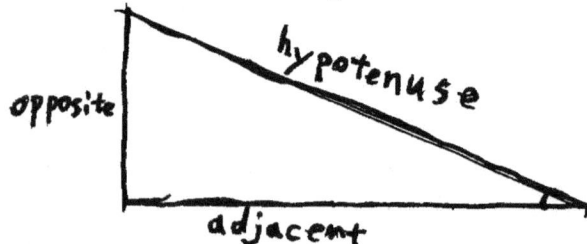

There are three main trigonometric function called sine, cosine, and tangent. The Sine of on angle (abbreviated as Sinθ, θ being a Greek letter called Theta) represents the ratio between the opposite side and the   hypotenuse (o/h). The Cosine of the angle (Cosθ) is the ratio between the adjacent side and the hypotenuse (a/h). The Tangent of the angle (Tanθ) is the ratio between the opposite side and the adjacent side (o/a), also referred to as "the rise over the run".

The sine and tangent of a 0° angle would be 0 since as a straight line there would be no opposite side. The cosine of that angle is considered to be 1 with the adjacent side & the hypotenuse merged.

For a 1° angle, the sine function is 0.01745. This can be figured using a scientific calculator or by drawing a right triangle with a 1° angle and comparing the measurements. Another option is to keep a function table on hand. Since the sine function is the relationship between the opposite side and the hypotenuse (o/h), a 1° angle with an opposite side of 100 centimeters would have a hypotenuse of about 5730.7 centimeters (100/ 0.01745). The cosine function of that angle is 0.9998, and the tangent is 0.01746 .

Here's a function table:

∠	sin	cos	tan	∠	sin	cos	tan	∠	sin	cos	tan
1°	0.0175	0.9998	0.0175	31°	0.5150	0.8572	0.6009	61°	0.8746	0.4848	1.8040
2°	0.0349	0.9994	0.0349	32°	0.5299	0.8480	0.6249	62°	0.8829	0.4695	1.8810
3°	0.0523	0.9986	0.0524	33°	0.5446	0.8387	0.6494	63°	0.8910	0.4540	1.9630
4°	0.0698	0.9976	0.0699	34°	0.5592	0.8290	0.6745	64°	0.8988	0.4384	2.0500
5°	0.0872	0.9962	0.0875	35°	0.5736	0.8192	0.7002	65°	0.9063	0.4226	2.1440
6°	0.1045	0.9945	0.1051	36°	0.5878	0.8090	0.7265	66°	0.9135	0.4067	2.2460
7°	0.1219	0.9925	0.1228	37°	0.6018	0.7986	0.7536	67°	0.9205	0.3907	2.3560
8°	0.1392	0.9903	0.1405	38°	0.6157	0.7880	0.7813	68°	0.9272	0.3746	2.4750
9°	0.1564	0.9877	0.1584	39°	0.6293	0.7771	0.8098	69°	0.9336	0.3584	2.6050
10°	0.1736	0.9848	0.1763	40°	0.6428	0.7660	0.8391	70°	0.9397	0.3420	2.7480
11°	0.1908	0.9816	0.1944	41°	0.6561	0.7547	0.8693	71°	0.9455	0.3256	2.9040
12°	0.2079	0.9781	0.2126	42°	0.6691	0.7431	0.9004	72°	0.9511	0.3090	3.0780
13°	0.2250	0.9744	0.2309	43°	0.6820	0.7314	0.9325	73°	0.9563	0.2924	3.2710
14°	0.2419	0.9703	0.2493	44°	0.6947	0.7193	0.9657	74°	0.9613	0.2756	3.4870
15°	0.2	0.9	0.26	45°	0.70	0.70	1.0	75°	0.96	0.25	3.732

	588	659	79		71	71	000		59	88	0
16°	0.2756	0.9613	0.2867	46°	0.7193	0.6947	1.0360	76°	0.9703	0.2419	4.0110
17°	0.2924	0.9563	0.3057	47°	0.7314	0.6820	1.0720	77°	0.9744	0.2250	4.3320
18°	0.3090	0.9511	0.3249	48°	0.7431	0.6691	1.1110	78°	0.9781	0.2079	4.7050
19°	0.3256	0.9455	0.3443	49°	0.7547	0.6561	1.1500	79°	0.9816	0.1908	5.1450
20°	0.3420	0.9397	0.3640	50°	0.7660	0.6428	1.1920	80°	0.9848	0.1736	5.6710
21°	0.3584	0.9336	0.3839	51°	0.7771	0.6293	1.2350	81°	0.9877	0.1564	6.3140
22°	0.3746	0.9272	0.4040	52°	0.7880	0.6157	1.2800	82°	0.9903	0.1392	7.1150
23°	0.3907	0.9205	0.4245	53°	0.7986	0.6018	1.3270	83°	0.9925	0.1219	8.1440
24°	0.4067	0.9135	0.4452	54°	0.8090	0.5878	1.3760	84°	0.9945	0.1045	9.5140
25°	0.4226	0.9063	0.4663	55°	0.8192	0.5736	1.4280	85°	0.9962	0.0872	11.4300
26°	0.4384	0.8988	0.4877	56°	0.8290	0.5592	1.4830	86°	0.9976	0.0698	14.3000
27°	0.4540	0.8910	0.5095	57°	0.8387	0.5446	1.5400	87°	0.9986	0.0523	19.0800
28°	0.4695	0.8829	0.5317	58°	0.8480	0.5299	1.6000	88°	0.9994	0.0349	28.6400
29°	0.4848	0.8746	0.5543	59°	0.8572	0.5150	1.6640	89°	0.9998	0.0175	57.2900
30°	0.5000	0.8660	0.5774	60°	0.8660	0.5000	1.7320	90°	1.0000	0.0000	∞

So if you see an object off in the distance that's straight ahead of you (0°), and you want to know how far away it is, first mark the position where you're standing with a stick or a rock or by scuffing the ground with your feet. Next, standing in that position and holding your arms out straight to the sides(~90°) while facing that object, if you look down one of your arms you can pick a position to move to. If you know you shoe size you can use that to measure the distance to the new position. Now holding out your arm you can count how many fist widths there are between the marker at your old position and the distant object. If there are 7 fist widths that makes it about 70°. Now you've made it into a right triangle and can easily figure out the remaining angle (180°-90°=90°, 90°-70°=30°).

If the distance to your old position (adjacent side) was 6 feet (180 cm), you can multiply that distance by the tangent of your angle (tan 70°=2.748) to get the distance from your old position to the object   (opposite side = ~16.5 feet(~495 cm)), or you can divide it by the cosine (cos 70°=0.3420) to get the distance from your current position to the   object (hypotenuse = ~17.5 feet(~526 cm).

Now suppose you wanted to measure the height of an object that was far away and high in the sky. Without having paper and a ruler to draft out the scale of the situation it's a little harder to make a right triangle to work with. Fortunately there is a way to use these functions with triangles that don't have a 90° angle.

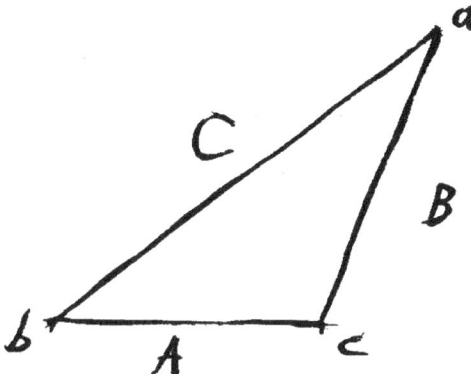

The ratio between the length of a side and the sine function of the opposite angle(referring to the ratio between the opposite side and hypotenuse would have if that angle were part of a right triangle) is the same as the ratio between either of the other sides and their opposing angles.

As a formula this is represented as:

a/sin A=b/sin B=c/sin C

Also, the ratio between two sides is the same as the ratio between the sine functions of their opposing angles.

a/b=sin A/sin B

These relationships are called the Law of Sines.

So if you saw an object ahead of you in the sky or at a significant height, first count how many fist widths it is above the ground. If it were 4 ½ fist widths, it would be ~45° up. Next, walk toward it a certain distance, in this case let's make it 10 feet [3 meters], and check the angle again. If the new angle is 50°, you can easily calculate the angle between the object and your previous location (180° straight line minus 50° equals 130°), and the angle between these two points as measured from the object(50°-45°=5°=180°-130°-45°).

The ratio of the distance from your old position to your new position and the sine of the angle between them is the same as the ratio of the distance between your new position and the object and the sine of the angle at your previous position.

10 feet/(sin 5°=0.08716)=~115=Current distance to object/ (sin 45°=0.7071). So the current distance to the object equals 115*0.7071=~81 feet [~24.3 meters]. Now that you have the straight line distance to the object [hypotenuse] you can figure the distance from your new position to a point directly underneath it and it's height above that.

There's another method of calculating the length of one of the sides when you know the opposite angle and the length of the other two sides called the Law of Cosines. It may come in handy when measuring an area.

$c^2=a^2+b^2-2ab*\cos C$

Okay, enough about triangles. On to...

Dealing With Other Shapes:

Other 2 dimensional shapes, no matter how complex, can be broken down and divided up into some variation or combination of those 3 simple shapes.

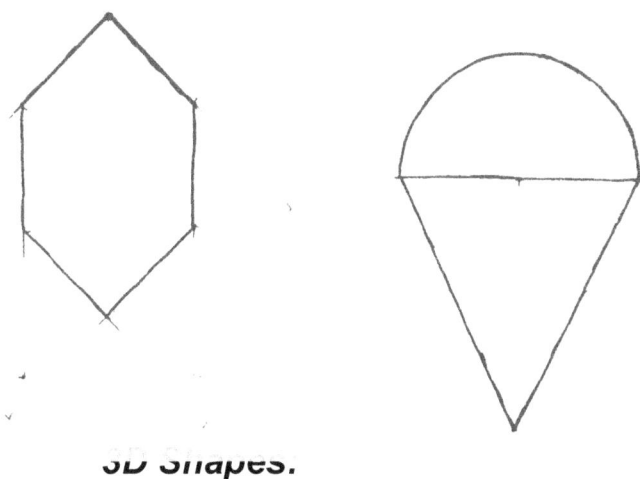

3D Shapes.

Cubes & Boxes:

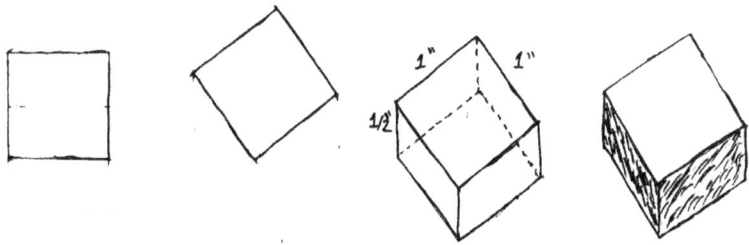

To draw a cube, it helps to start with a square. Turn it a bit so the corner is facing toward you and draw another one a bit above; below; or to the side of it. Then connect the corners. You can use dashed lines for the ones on the other side, or you can hide them and use shading to indicate a light source and shadows.

This is a diagonal Perspective drawing, so even though the sides of the object may be the same size, depending on the angle you look at them, they may appear to be longer or shorter. You could also keep everything the same size as the real object and adjust the angles to make it fit in an Isometric drawing. Of course many boxes aren't cubes. They may be based on rectangles or other shapes. A box with a circular base is called a Cylinder. A box with a triangular base is called a Prism shape.

To find the surface area of a box or a cube, add the surface areas of each of it's surfaces. For a cylinder, add the surface area of the top and bottom circles and multiply the circumference by the height for the outside surface.

To find the interior area or Volume of a cube or rectangular box multiply the length times the width; times the height. For cubes these are all the same number, so you're multiplying a number by itself and multiplying that product by that number again, a process appropriately known as Cubing. If the dimensions were measured in feet the volume is expressed as cubic feet or feet cubed [ft^3]. Centimeters become centimeters cubed [cm^3] or cubic centimeters [cc]. For fluids 1 cubic centimeter equal 1 milliliter [ml]. 1 gallon=~231 in^3 & 1 fluid ounce [fl. oz.]=~1.8046 in^3.

For cylinders and prisms, multiply the surface area of the base circle or triangle by the other dimension.

Now that we are in three dimensions there a couple of important new calculations, specifically the density and the center of balance for the object. The Density or how dense an object is refers to how concentrated its weight or mass is. Just divide the weight or mass by the volume.

For an object with a simple shape and a uniform [meaning the same throughout or all "one form"] consistency [meaning what something consists of or is made of] the Center of Balance is the point where the center point of all the surfaces meet. It the object is tipped so that this point has passed the base, the object will fall over. That's also called the Tipping point.

Spheres:

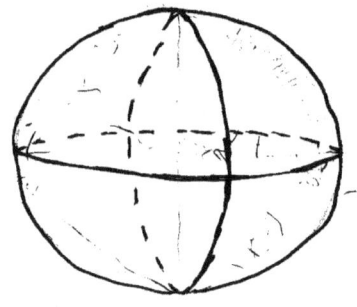

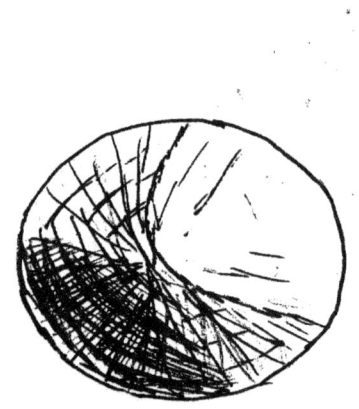

To draw a sphere, first draw a circle, then draw lines like the slit of a cat's eye on either side of the horizontal and vertical center-lines. Make one of those near the horizontal center and one near the vertical center into dashed lines to represent the opposite side. The other method is to use shading indicating illumination and shadow to create the effect of depth.

The surface area of a sphere is about four times that of a circle with the same radius. The formula is: $A=4\pi R^2$

The volume of that sphere is equal to 1/3 of the surface area multiplied by the radius. That formula is: $V=4/3*\pi R^3$

Spheres or globes are often used as models and maps of planets. For the planet Earth we use two sets of numbers, called latitude and longitude, to indicate specific points on our planet.

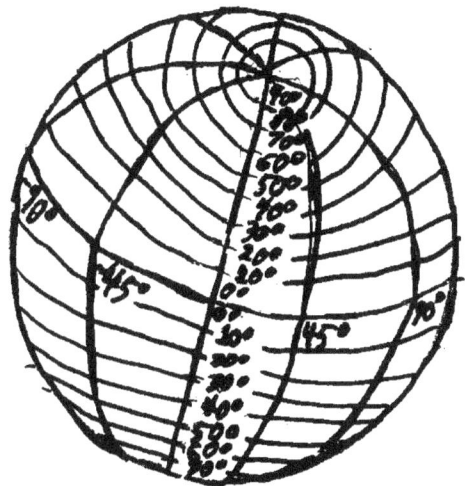

Latitude refers to the horizontal lines or circles on a map or a globe. The middle line or circle drawn on a map or globe of the Earth is called the Equator and has a numerical value of 0°. The lines above and below it   increase in value the further up [North] or down [South] they are, up to 90° for the points at the polar ends.

Longitude refers to the vertical lines that are drawn from pole to pole. The line with the value of 0° ,commonly called the Prime Meridian, runs through a town called Greenwich in the country currently called the United Kingdom and continues south through part of the continent of Africa. Longitude lines east of the Prime meridian have a positive degree value while lines to the west are negative. Opposite the Prime meridian at 180° is the Anti-meridian, roughly marking the official international date line where each new calendar day officially starts at 12:00am.

For more accurate coordinates, the space between each degree of longitude & latitude is further divided into 60 segments called minutes['], which are in turn divided into 60 segments called seconds["]. A degree of latitude is approximately 69 miles, and a minute of latitude is approximately 1.15 miles. A second of latitude is approximately 0.02 miles, or just over 100 feet [30 meters].

A degree of longitude varies in size. At the equator, it is approximately 69 miles, the same size as a degree of latitude. The size gradually decreases to zero as the meridians converge at the poles. At a latitude of 45 degrees, a degree of longitude is approximately 49 miles. Because a degree of longitude varies in size, minutes and seconds of longitude also vary, decreasing in size towards the poles.

Pyramids & Cones:

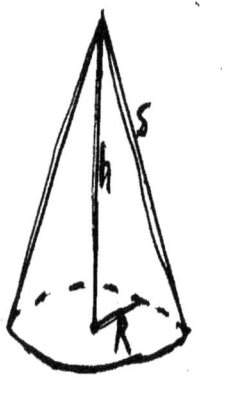 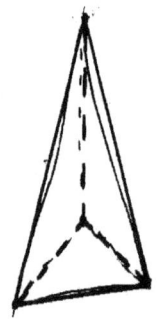 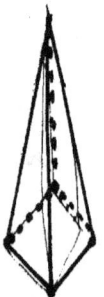

The surface area for a cone is equal to the area of its circle [πR^2] added to the outer surface of the cone. The area of the outer surface is equal to the ½ of the length of the side multiplied by the circumference of the circle [$1/2s*2\pi R$]. The formula is: $a = \pi R^2 + s\,\pi R$

The volume of a cone or pyramid equals 1/3 of the height multiplied by the area of the base shape.

If you want to build your skills and turn simple 2 dimensional shapes into complex shapes 2 & 3 dimensional shapes, you might want to study and practice origami, the art of paper folding.

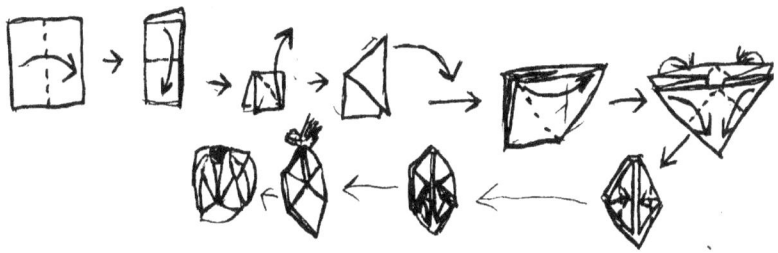

5 - Working with Tools:

Starting with:

Ramps & Wedges:

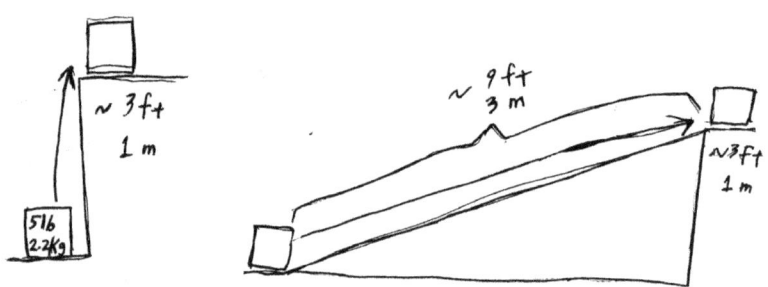

If you need to raise an object off the ground, lifting it directly you'll be using a little more force than the weight of gravity on its mass. The energy needed to perform that Work is usually calculated by multiplying that force by the distance the object is moved.

For the Old English system a 5 pound object being moved 3 feet up would take 15 Foot*pounds of energy. In the systems international or metric system, the same object, having a mass of 2.2 kilograms is being lifted against the acceleration of Earth's gravity (9.8 m/sec^2) for a force of 21.56 Newtons. To be lifted 1 meter that object requires 21.56 Newton*meters or you could also say 21.56 Joules of energy.

By pushing or pulling that object up a ramp, you're still doing the same amount of work, but you reduce the amount of force required. If you use a ramp that's over 9 feet (~3 meters) to raise an object a little over 3 feet (~1 meter) the Ideal Mechanical Advantage would be about 3 (distance to be moved ÷ height to be raised), which means the amount of force required would at best be divided by 3, making it 1$^2/_3$ of a pound or about 7.19 Newtons to move that 2.2 kg object. If you know the angle of the ramp from the underlying surface, the ideal mechanical advantage is the reciprocal of the tangent[opposite side ÷ adjacent side] of that angle.

To be more accurate, the Work done by a force is changing a vector, which could be a short movement in a given direction, or changing the speed or direction something is moving.

A Vector is an amount with a direction. A common example of a vector is Velocity which is the speed in a given direction.

The amount of work done is equal to the portion of the force that is in line with the with the new vector[F cosθ] multiplied by the vector change[v].

You could abbreviate the formula as W=Fv cosθ.

Due to the law of conservation of energy, when a force does work on an object, it changes the energy of the object by a similar amount. The 15 foot*pounds or 21.56 Joules of energy is now considered to be part or the object's Potential Energy. As long as it's not moving on its own or being used this can be called Static Energy. If the object is allowed to fall, or if its weight is used to move something else, that becomes Kinetic Energy [kinetic means motion].

Since there is going to be some friction or resistive force between the object and the surface of the ramp, the actual force required might be around 2 pounds (8.9 Newtons), so the Actual Mechanical Advantage would be 1½ (5 pounds of weight ÷ 2 pounds of force). The Efficiency or the comparison of the work done to the energy required to do it would be 50% (AMA/IMA=1.5÷3=0.5x100%=50%).

The efficiency and mechanical advantages of a wedge used for splitting logs or holding objects apart can be figured the same way.

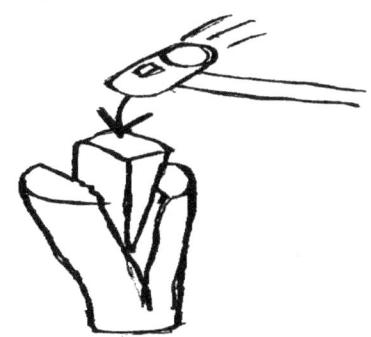

To make a ramp you just need a board or platform to extend diagonally from what ever height to the floor or ground. To make a Wedge you need an object shaped like a triangular box or a pyramid. A variation on the basic wedge is a chisel. Depending on your application it could be made of wood; metal; stone; or clay.

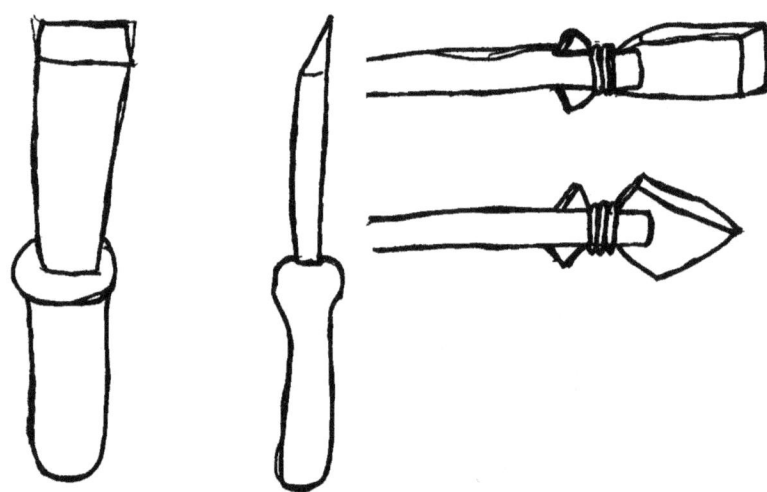

There are different types and shapes of chisels. Sharp chisels intended for soft material like wood can be used by simply pushing through the   material, cutting it and gouging it out. It's a good idea to cut an outline before chiselling out the material to be removed.

If you want to remove the material in an even line without going deeper, you should keep the angled side of the chisel down.

A carpenter's plane for smoothing out rough wood works in much the same way.

There are some chisels, especially those from Japan, that are curved in on the opposite side to reduce friction.

Chisels intended for carving hard materials like stone are designed to be struck on the butt of the handle with a hammer. In the case of stone, the first strike stresses the material, causing small or even microscopic [too small to see with the average naked eye] cracks. Striking it nearby at a different angle can send the material between the two points flying. Just like with softer materials, it's still a good idea to cut out an outline before you try to remove larger sections. The material near a corner is easy to remove, and this can be done even using softer materials like wood or antlers from an animal.

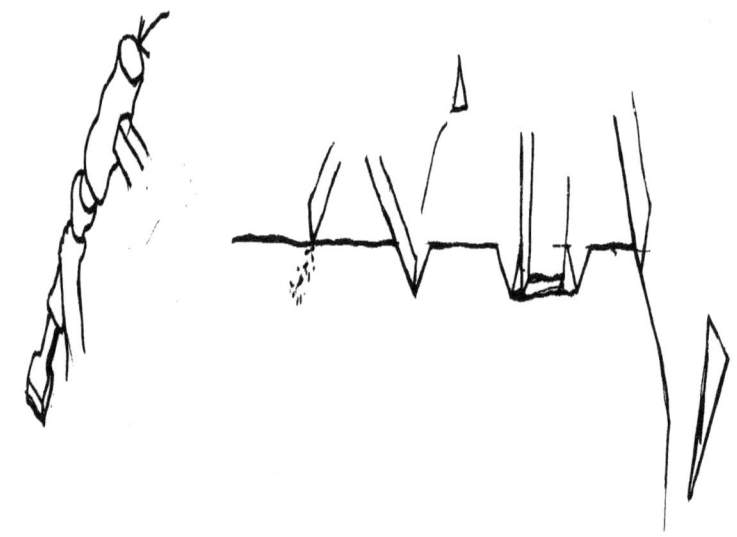

If you have multiple wedges sticking out of an object, it can be used to remove a lot of material at once. An example of this would be a saw or a rasp. A Rasp is a type of filing tool with a very rough surface. The bigger and further apart the wedges the more material they remove. The smaller and closer the are, the smoother the surface will be. This grain is often represented by the number per square inch or square centimeter.

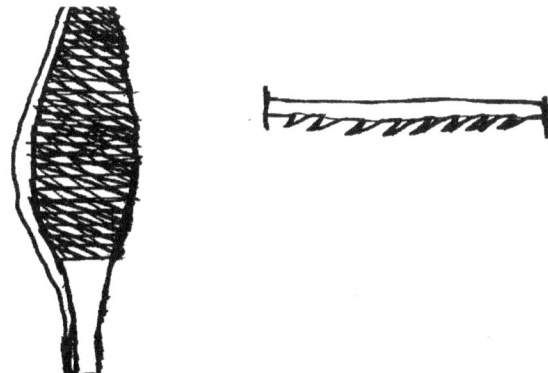

Another application for a wedge is as an awl [hole making tool], or with a bit of creativity, even a rotary drill can be made with simple materials.

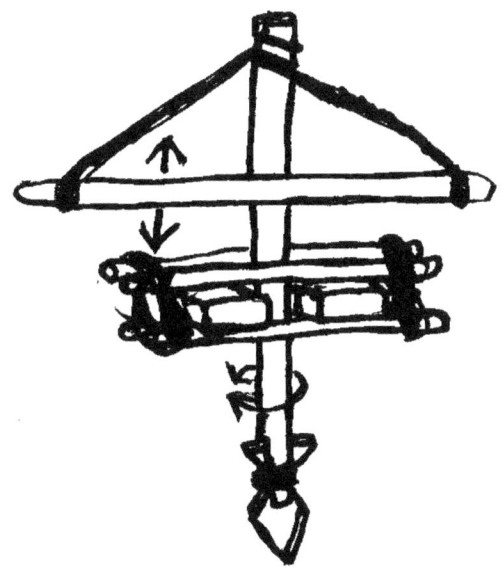

The next type of tools are:

Levers:

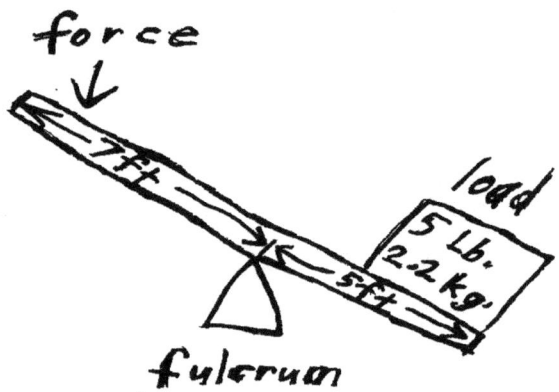

A lever consists of 3 parts that may be separate or part of each other:

1) an arm or bar for applying force,

2) a connection with a load object to be moved or manipulated, &

3) a fulcrum that redirects the force against the ground or some other point of contact.

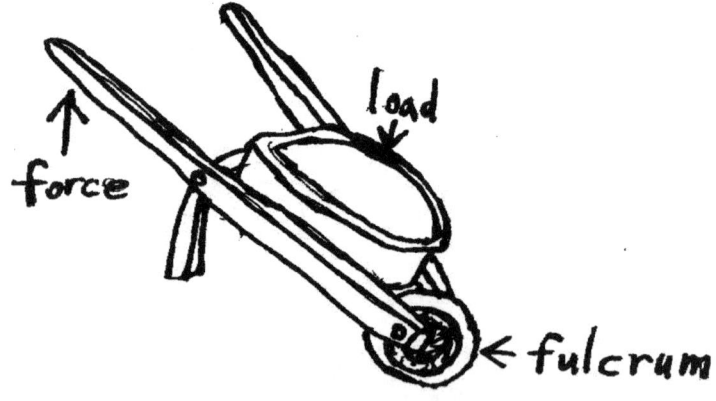

Depending on the arrangement of these there are 3 different classes of levers. A Class 1 lever has the fulcrum in between the force and the load. A Class 2 lever has the load between the force and the fulcrum. A Class 3 lever has the force between the fulcrum and the load.

The ideal mechanical advantage of a lever is based on the ratio of the distance of the force from the fulcrum and the distance of the load from the fulcrum. The further away the force is, and the closer the load is, the greater the mechanical advantage. There are some tools, especially class 3 levers, that have a mechanical advantage that's less than 1. These are intended to either increase your control, for manipulating small or   delicate objects, or to increase the speed and distance the load is moving through. One example of this is how an axe or hammer is used.

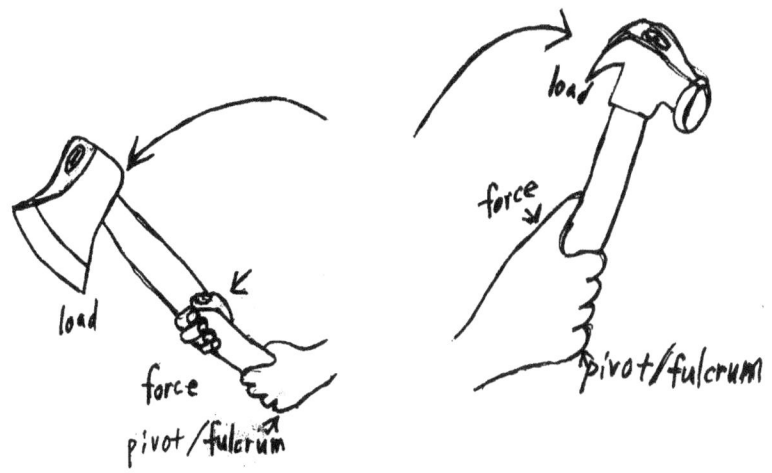

These two are some of the earliest manufactured tools. If you have a branch that naturally splits in a Y or has some kind of split in it, you can easily insert an appropriately shaped stone, or even a piece of clay that has been moulded & fired, and lash it into place.

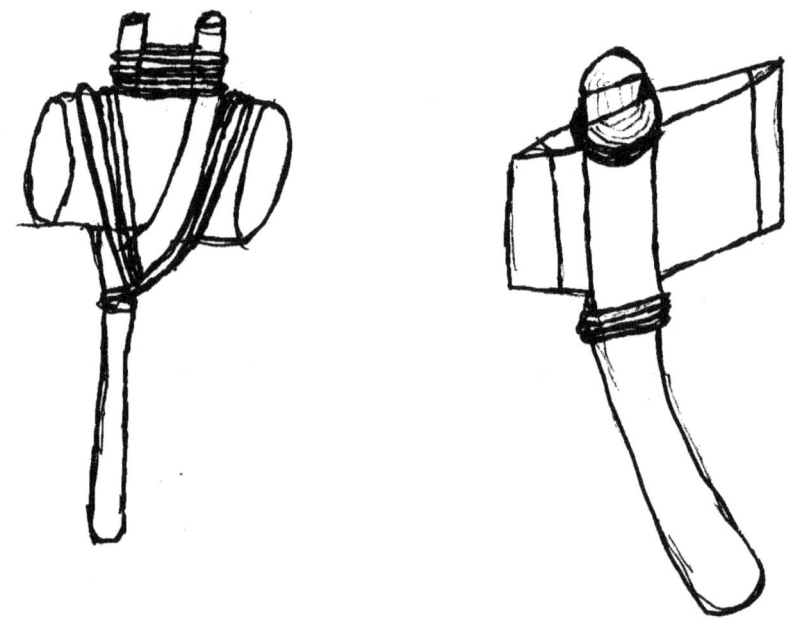

Usually the friction on a lever's fulcrum is small, making the actual mechanical advantage closer to the ideal mechanical advantage and increasing efficiency. By combining levers and wedges we get excellent tools for shaping materials like the axe or hatchet or for removing obstacles like these class 2 levers.

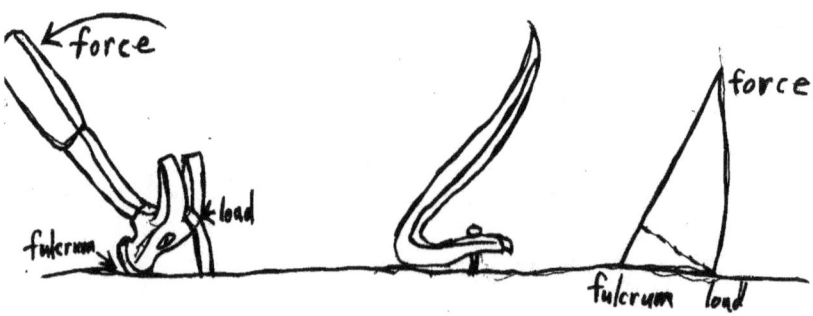

We can also make a double armed lever and pin the arms together at the fulcrum for a pair of tongs.

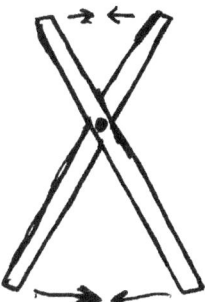

This allows us to pinch and hold things for easy manipulation or scissor them.

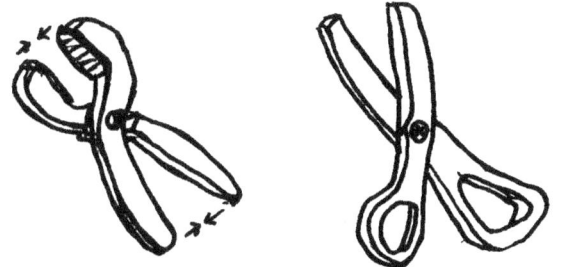

All levers can be said to exert a rotational force called Torque.

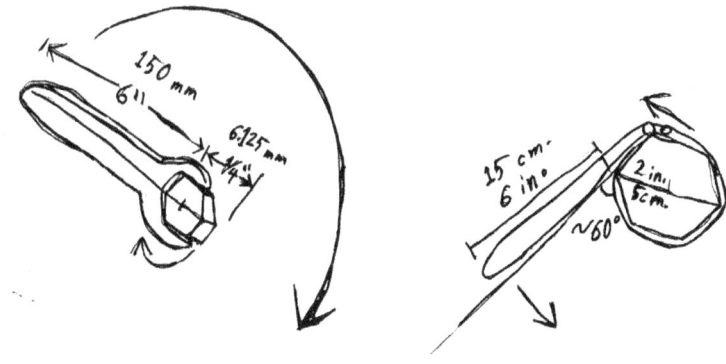

The relative torque is the same for both the force side and the load side and for a straight lever is equal to the force or load weight multiplied by its distance to the fulcrum or pivot point. Since we're talking about potentially rotating in a circle, that distance can be called the radius.

For a lever that is bent or set at an angle to the load like many wrenches are, especially those designed to fit a variety of load sizes like an oil filter wrench, the torque is equal to the cosign of the angle times the radius times the linear force applied. Of course, if it's straight; the angle is 0° & the cosine [adjacent/hypotenuse] is 1, which doesn't affect the product. So the formula is abbreviated as:

Torque = $rF(\cos\theta)$

Even though this is a force times a distance, the product is still in force units. The reason for that is the distance is not how far the object has moved. It's just a limiting or magnifying factor for the force.

Now that we've discussed torque we're ready to look at:

Pulleys, Gears & Other Rollers:

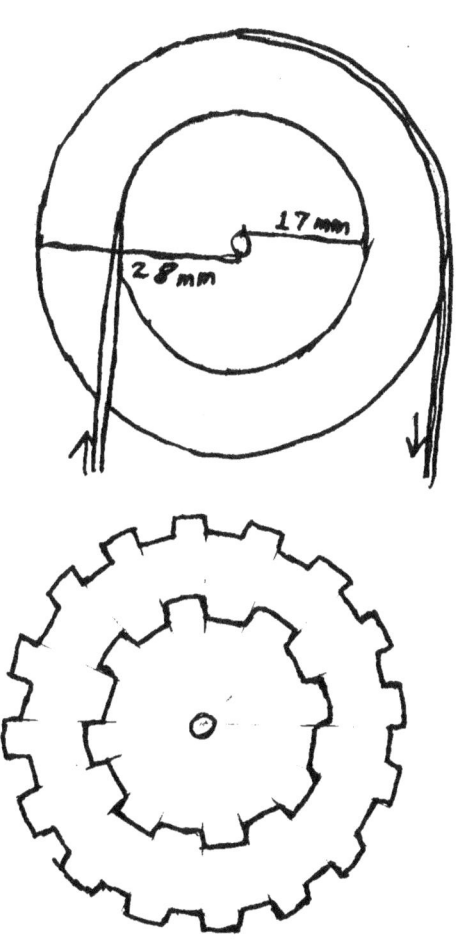

A pulley is basically a wheel with a groove for belt or rope and a gear is a wheel with teeth for a better transfer of force. As you learned with levers the farther away from the center a force or a weight is, the greater its rotational force or torque is.

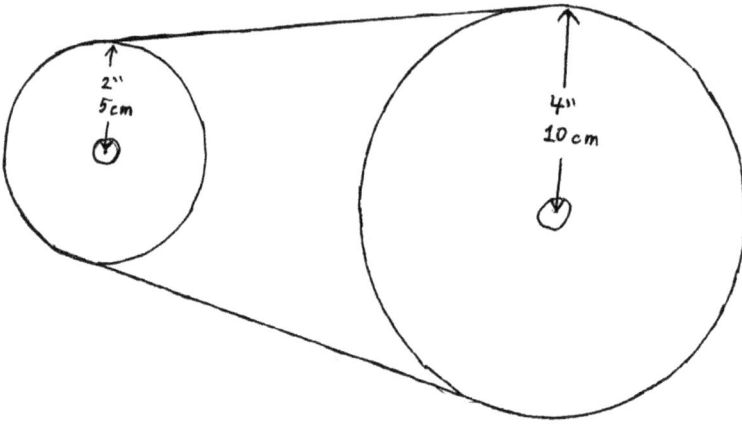

So for two pulleys that are belted together, two gears with their teeth meshing, or two pulleys or gears mounted on the same axle, you have an ideal mechanical advantage for one turning the other based on their   relative sizes. Just think of the radius being the distance from a lever's fulcrum. For the gears you can also figure it out based on the relative numbers of teeth.

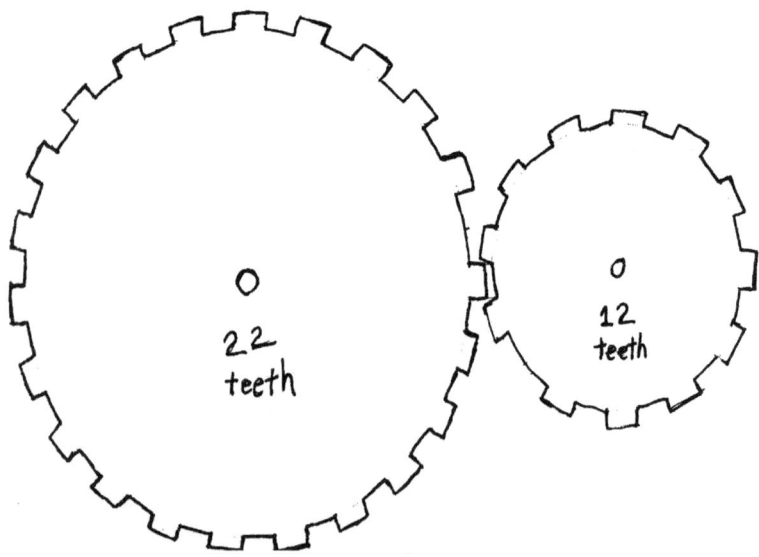

For a pulley lifting system the pulleys are usually all the same size. The ideal mechanical advantage is equal to the number of rope sections that are supporting the weight. Of course even a single pulley can be useful in positioning an object or redirecting a force.

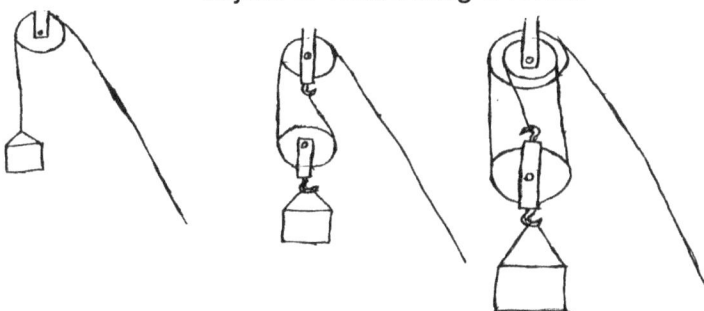

The friction from sliding one body over another can be called "Sliding friction". When an object is rolled over a surface, the resistance is greatly reduced and is called "Rolling friction". Since rolling friction takes up less energy, many mechanical systems use some kind of wheel; ball; or rolling cylinder to minimize rubbing and reduce the wearing down and damage that can occur.

Pulleys, gears, and other wheels can be cut, moulded, or otherwise shaped from various materials including glass, ceramic clay, metal, plywood, etcetera. Straight grained wood often has a problem of splitting from the torque when used for pulleys or having the teeth break off for gears. Plastics often soften and pull away from the axle or slip off. If the gear is intended to spin freely the axle can be loose or mounted to minimize friction.

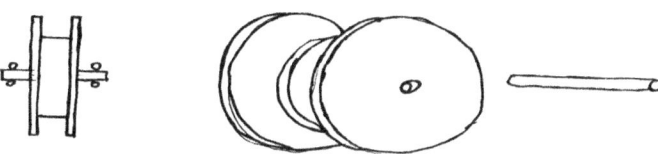

If it's intended to move a load on that axle or be turned by that axle, you'll need some way of securely attaching it to transfer the torque. Glues & epoxy don't stick as well to a smooth and round surface.

In order to more easily create round objects, there is a tool called a Lathe. The simplest form of this device is two post mounted at a fixed distance with a log suspended between them by pins coming out of the center-points. This can be rotated by pulling a chord that has been wrapped around it, letting it rewind with its rotational momentum. This back an forth rotation can be maintained with a foot lever a the spring action of a tall tree sapling [also called a lath, which is where the name of the tool came from]. Now the log or other material can be smoothly shaped with chisels and other tools.

With a few bearing wheels or balls and some lubricant, a horizontal turntable version can be made for working with clay.

Once a wheel starts turning it tends to keep turning and the bigger or heavier it is and the faster it turns, the longer it will keep spinning. There are a number of devices that use this Flywheel effect to improve their efficiency.

The tops used as toys all around the world operate on the same basic principles. As it spins it there is an effect where anything on the surface is slung to the perimeter with a force that increases the the faster the top spins and the bigger it is. This is called Centrifugal force.

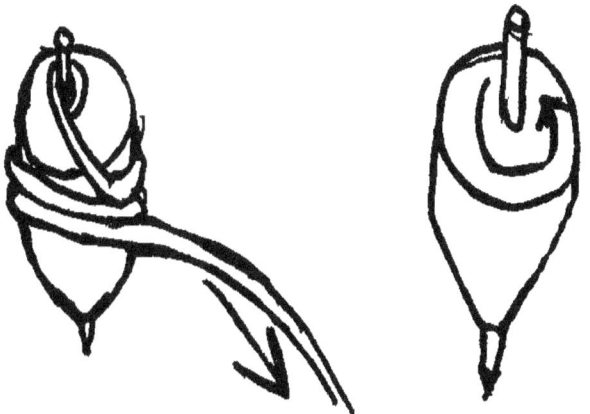

It also has a tendency to resist having its angle changed and will push back or away if tipped. Some people don't think centrifugal force sufficiently explains this, so the term "Gyroscopic force" has been developed. There is another device that has become popular as a toy called a Gyroscope. It has a flywheel mounted to turn freely on pegs attached to a metal ring.

Since the frame doesn't move with the wheel, but still shows the same tendency to resist tipping, the gyroscopic effective can be used to stabilize sailing and flying vessels. It has also been used to make directional compasses that are more reliable than magnetic compasses. On the other hand, for devices that are going to be moving and turning, the gyroscopic effect can cause damage to a flywheel, depending on the angle it's set at.

Adding Vectors:

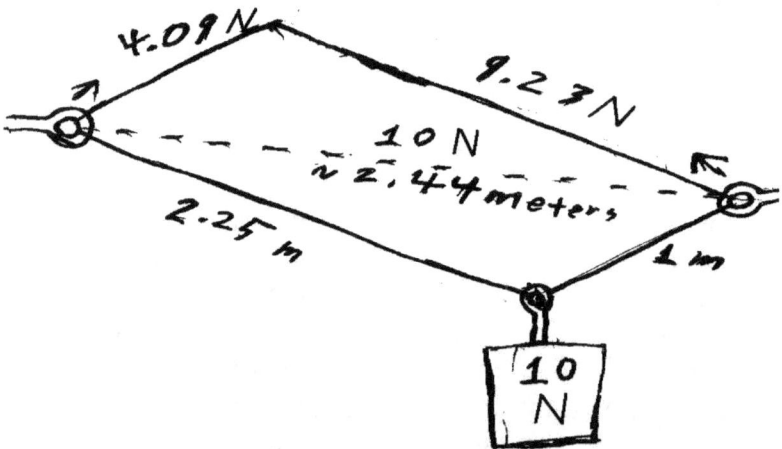

After all that, it may surprise you to find out the force on the fulcrum, axle, or pivot point isn't equal to the sum of the other forces. Since each of these forces is going in a specific direction, each one is considered a vector.

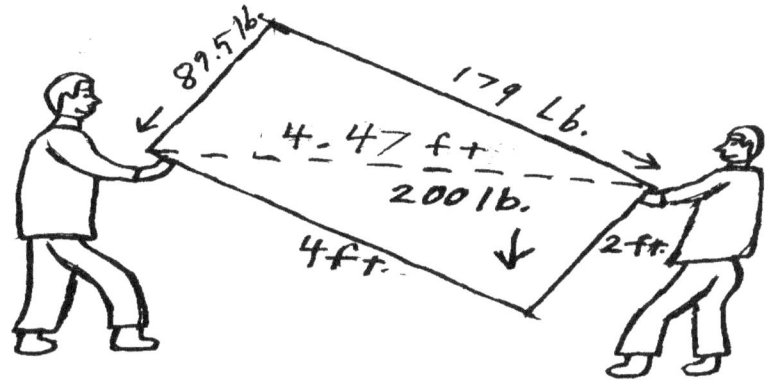

Adding Vectors is like adding the short sides of a right triangle to get the hypotenuse. The square of the resultant force on the fulcrum or axle is the sum of the squares of the vectors. The same thing happens when an object is supported or moved by more than one force. Also the force that's closer to the load or closer to the center of the weight is the one that takes more of the strain or most effectively adds to the resultant force.

Unlike the triangle's hypotenuse, there may be more than 2 vectors   contributing to the resultant force. Just find the square root of the sum of the vectors' squares.

$$_RF=\sqrt{_1V^2+_2V^2+_3V^2+...}$$

If you want to know why to resultant force is less than the sum of it's parts it helps to look at...

Graphs:

Vectors have a quantity and a direction. Taking velocity for example, suppose a car is travelling north at 40 kilometers per hour[kph]. Lets draw a vertical line and a horizontal line that cross in the center. Mark a 0 near the crossing point and a mark small line at every centimeter to represent 10 kph in each direction [north, east, south, & west]. This type is often called an XY graph because the numbers on the horizontal line are often listed as positive and negative units of an X axis and the vertical as the Y axis.

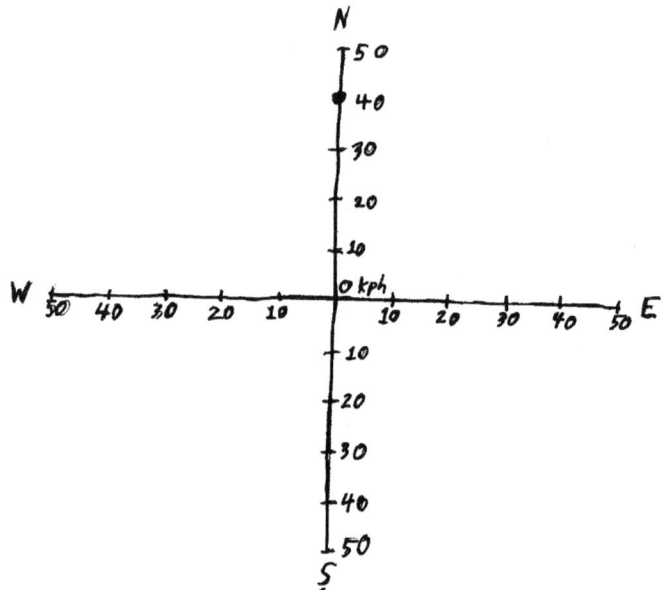

Now make a big dot at the 40 kph north point.

If this car were to turn 45° to the north west without changing their   forward speed, how would that look on this chart? Draw a line 45° to the left of the north from the 0 point and make a mark at 4 centimeters along this line.

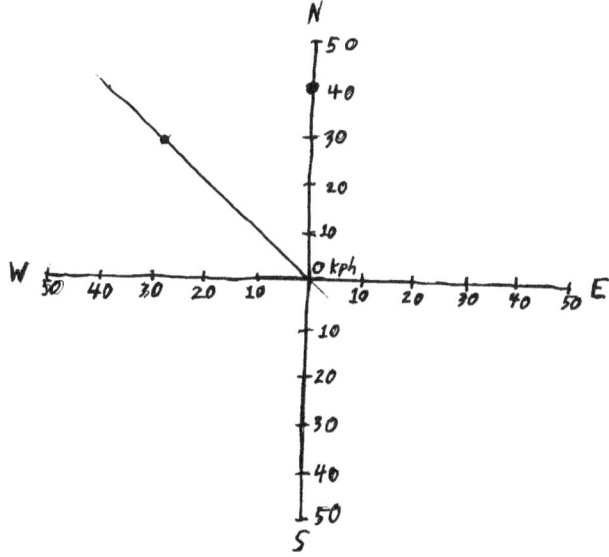

Now how fast is this car going north and how fast is it going west? Let's measure and find out.

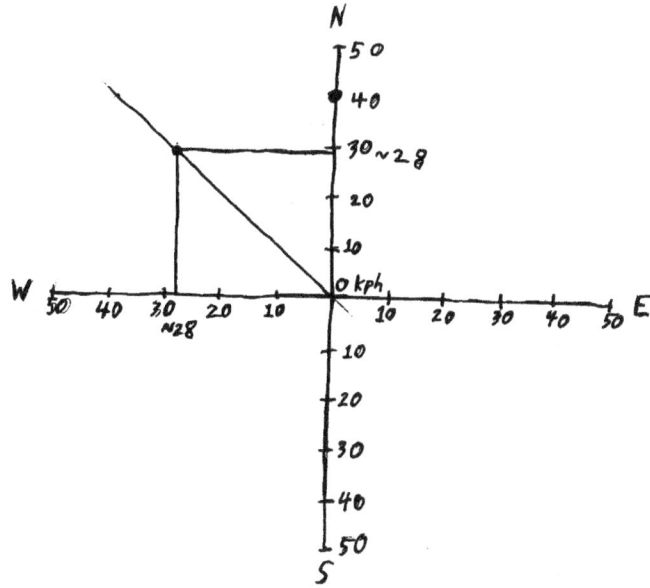

Well, the new point is 2.8 centimeters up along the northbound axis and the same distance left along the westbound axis, making the speed; relative to each of those directions; 28 kilometers per hour.

If we compare that to the formula for adding vectors: $\sqrt{(28)^2+(28)^2} = \sqrt{784+784} = \sqrt{1568} = $ ~39.6, that's pretty close.

Just by changing directions without changing forward speed, the velocity was changed by 12 kph to the north and 28 kph to the east for a total difference of 40 kilometers per hour. That's as much of a total change in velocity[acceleration] as if the car had come to a stop or doubled its speed.

That's why they say it's a bad idea to take sharp turns or curves at high speeds. Considering the inertia or momentum of a 1 ton vehicle, there may not be enough friction to keep the wheels on the road!

Now using force as our vector, let's look at a jet aircraft with 4 engines. Each engine is providing 100 pounds [445 Newtons] of thrust. If we run the formula for adding vectors $\sqrt{4(100)^2} = \sqrt{40000} = 200$ pounds! You might be asking "Where did all that force go?"

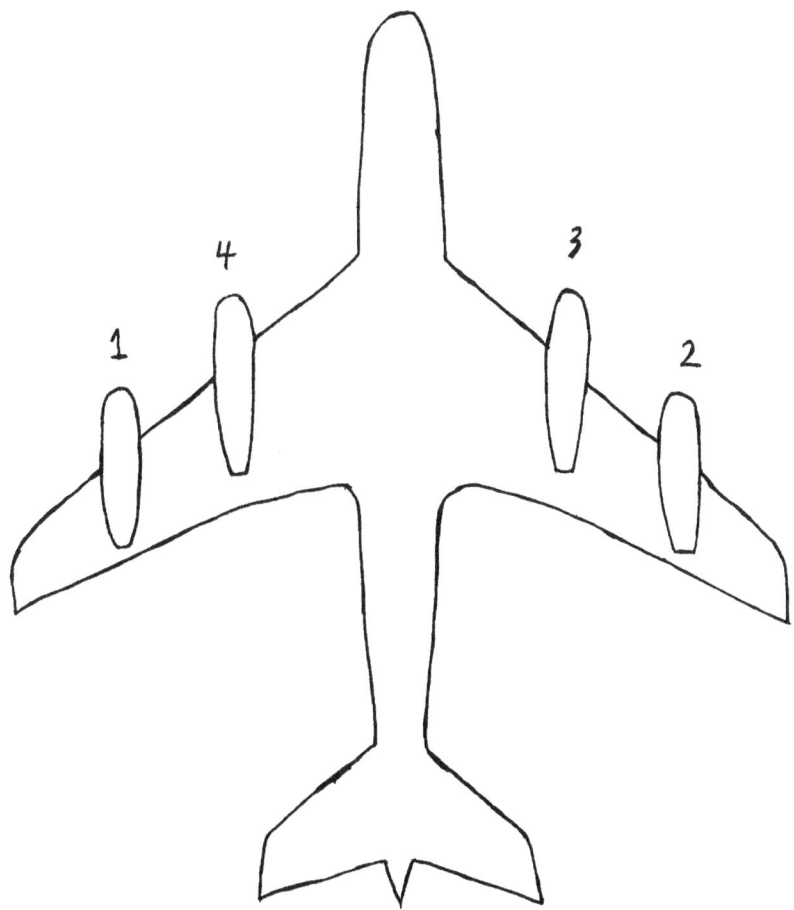

Since they are mounted on different parts of the plane; the angle each   engine would move the plane is different.

One way to think of this is, if the jet-plane were on a friction free surface and only one engine was firing, the point of the tail would move through the point occupied by the engine as the plane turned in a circle.

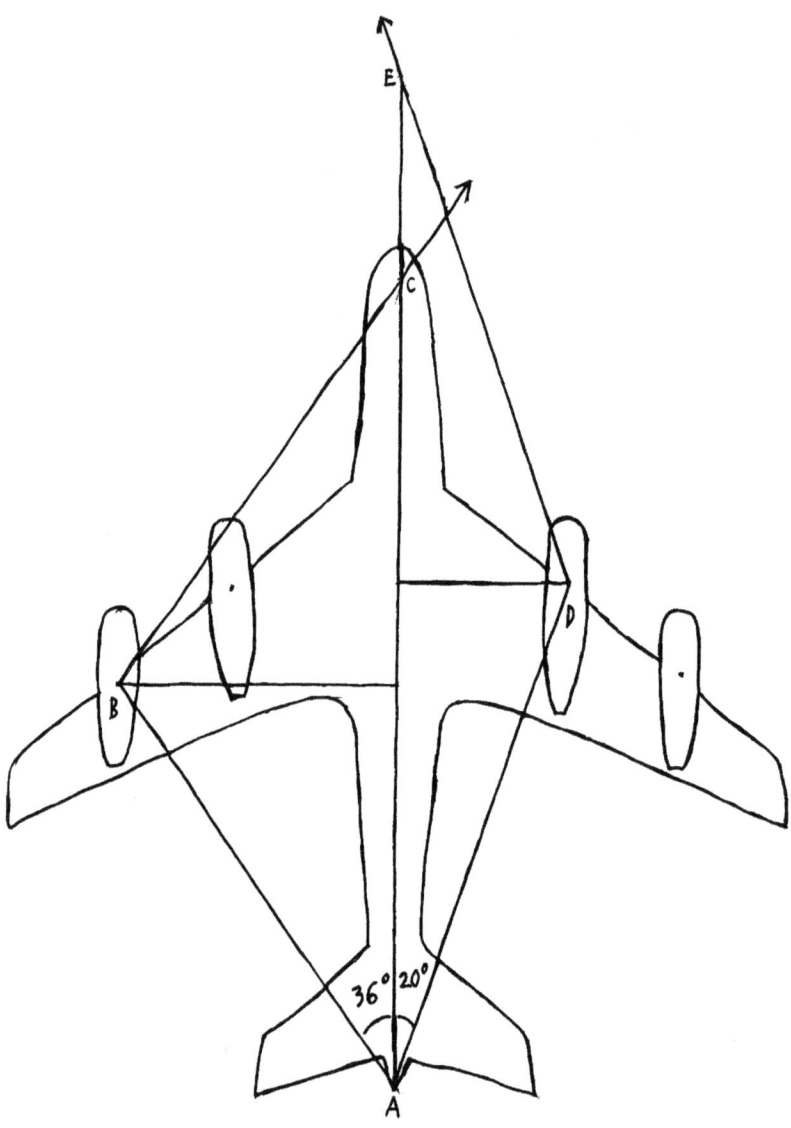

We can make a line graph to illustrate how engine 1's angle of rotation reduces its forward thrust. Graph paper makes this easier, but it doesn't show up very well on a black & white scan.

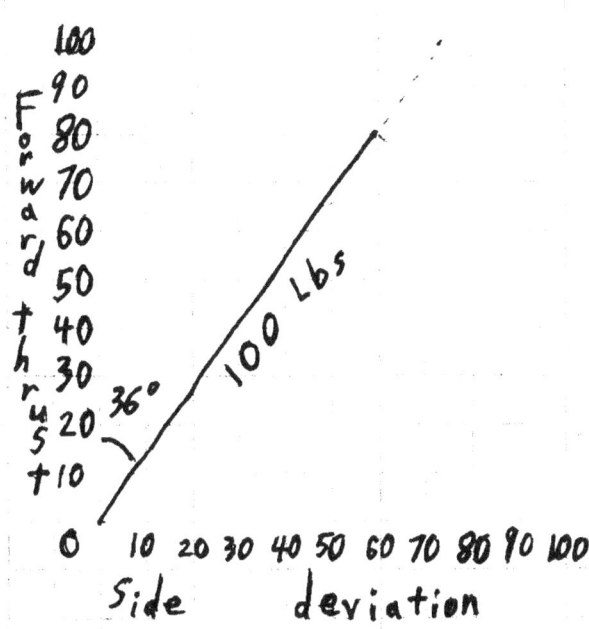

That's where part of the thrust went. Now let's see how much engine 2 cancels it out.

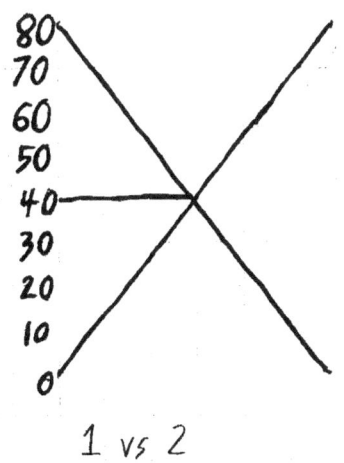

1 vs 2

Working against each other like this they're only about half as effective.

Let's look at the performance at engine 4's position.

Being closer to the center of the plane seems much more effective.

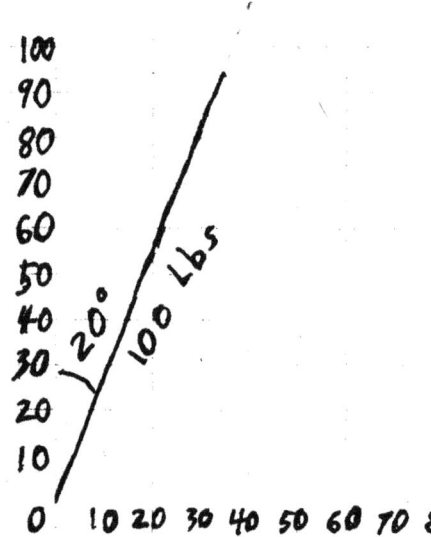

It's still losing some by working against its partner.

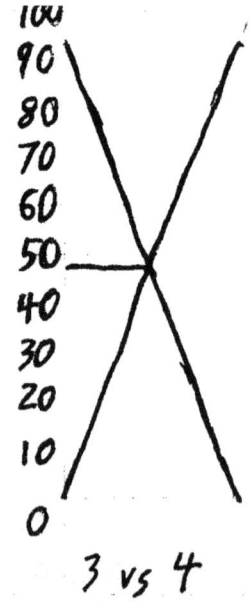

3 vs 4

We can also compare the force mismatch from engine 4 to engine 2.

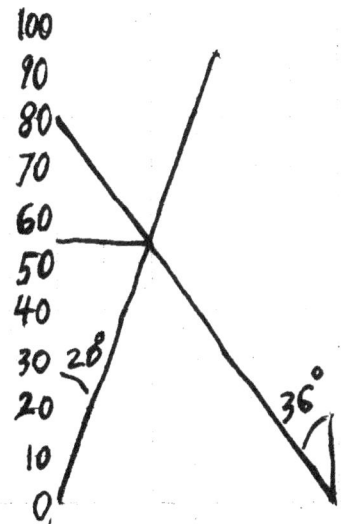

With them being off balance we get more forward thrust and some sideways drifting. Fortunately with all four it balances out.

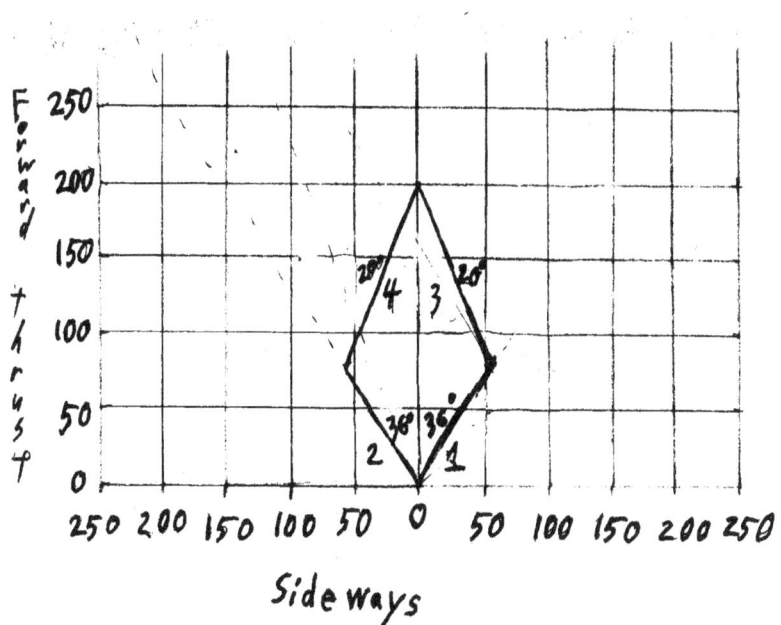

One of the reasons this happens is because the plane's wing acts like the arm of a lever.

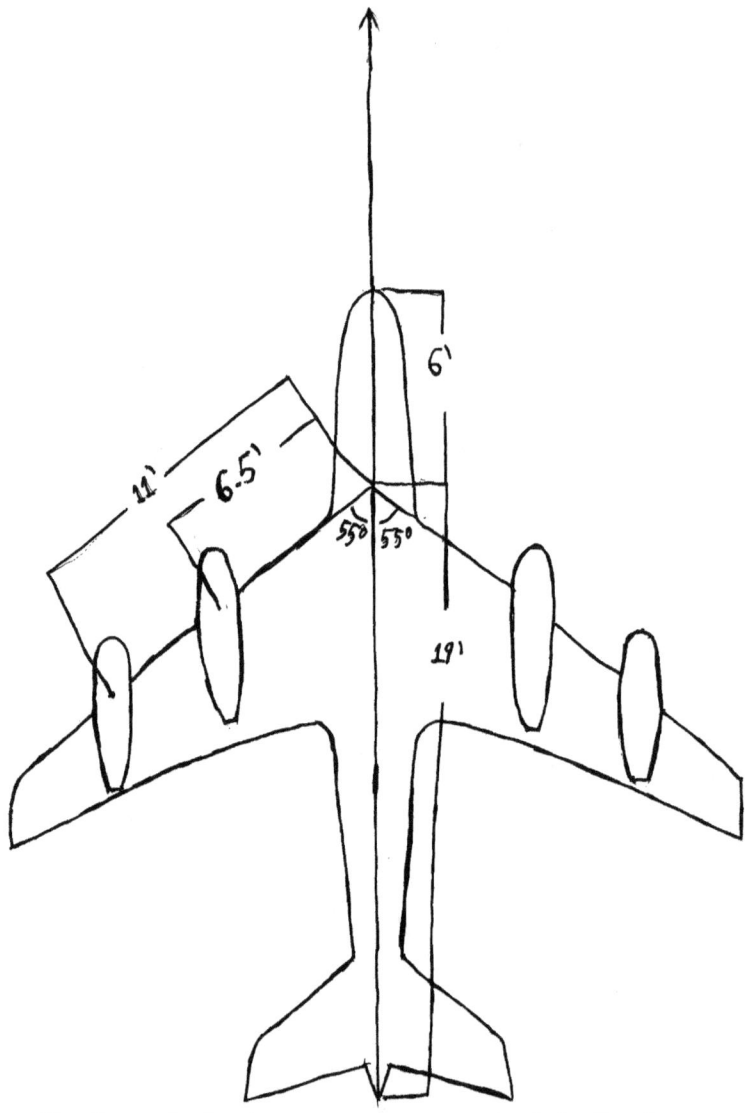

Using the formula for torque you can calculate the torque each engine exerts on the body of the plane.

For engine 1 & 2 that equals 100 lbs x 11'(cos 55°)=~631 ft*lb. On the 6 foot nose of this plane that's exerting ~105.2 pounds of force. For engines 3 & 4 that's ~344ft*lb and ~57.3 pounds against the nose. That information really stands out in a bar graph.

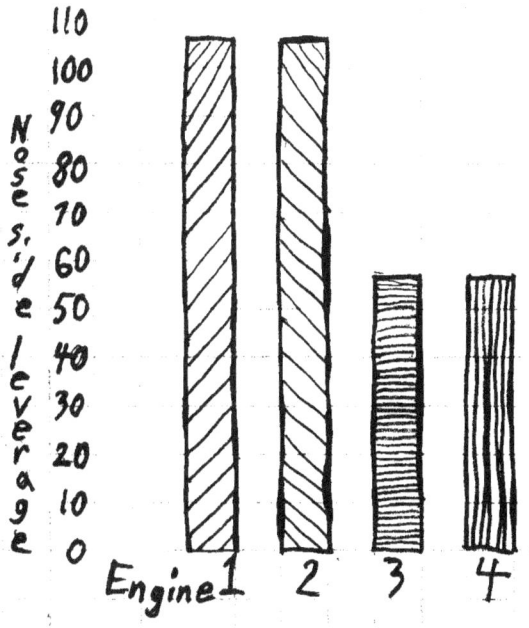

Finally we can get down to the

Screws & Bolts:

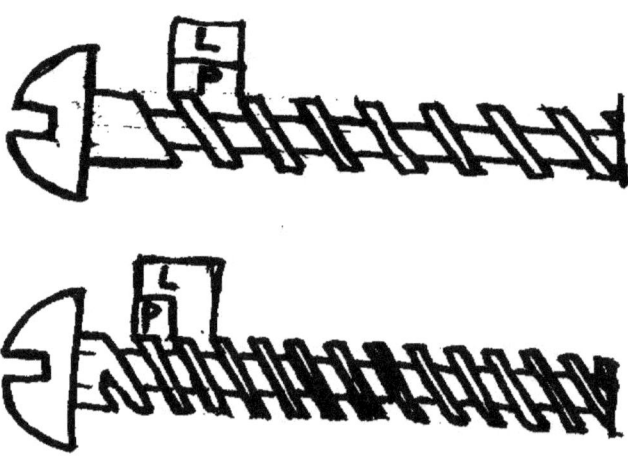

The ridges twisting along a screw's length are called threads. Other than the length and width, the important figures for screws are the Pitch and the Lead [pronounced "l-ee-d"]. The pitch is the distance from the front edge of the top of one thread to the front edge of the top of the next thread. The lead is the distance the screw will advance in a nut for each complete turn of the screw. In a single threaded screw, where there is only one line of threading cut around it, the pitch & lead are equal. In a double threaded screw the lead is twice the pitch, and in triple threaded screws the lead is three times the pitch. The pitch may be stated as a size per thread or as threads per inch in English system and threads per centimeter or millimeter in metric/SI.

Knowing the pitch & lead and setting up graduations for partial turns   allows us to make devices called Micrometers, that can very precisely and accurately make measurements in very small increments.

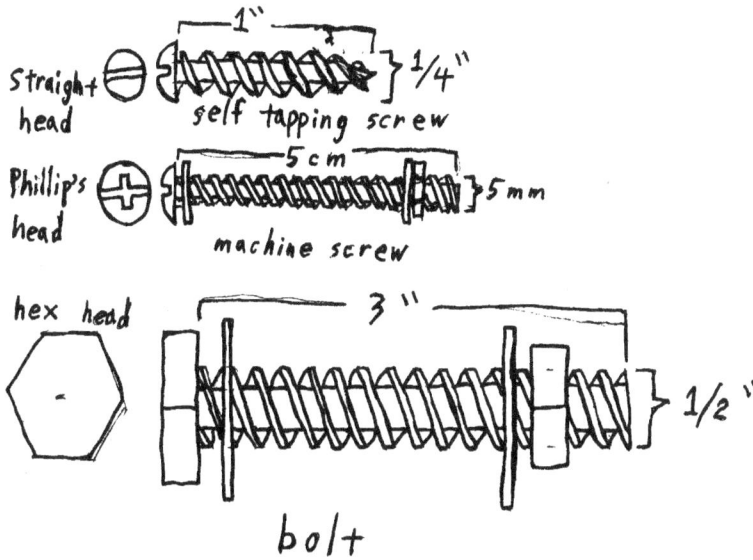

A Screw may be a self tapping screw with a point on the end and sharpened threads, a thin machine screw, or a thick bolt. It may have straight slot in the head for a regular screwdriver, a crossed slot for a Phillip's head screwdriver, a hex head for socket or a wrench, or various other designs.

A Nut is a block of metal or other material with a threaded hole for attaching to the threads of a screw.

A Washer is a disk or plate of metal or other material with a big enough hole through the center to insert a screw though it without touching the threads and small enough that the head of the screw won't go through it.

These basic parts in different sizes and with different attachments allow us to make adjustable tool like monkey wrenches, clamps, and vices.

A screw can be used to turn a gear in a combination called a Screw gear or a Worm gear. The screw [in this case often called a worm] turns the gear slowly. In theory you could use the gear [called, in this combination, a worm wheel] to turn the screw/worm very quickly, but the teeth need to be angled to match the threading to prevent damage and a well lubed bearing system is needed to prevent jamming in place.

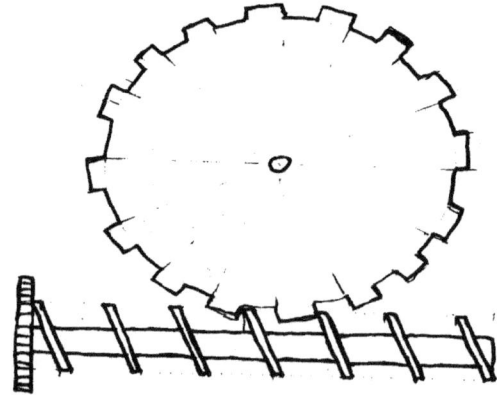

A Twist drill is designed like a self tapping screw with the sharp thread angled back away from the work to cut into the sides and move the   material out of the hole, but the tip is angled to cut into the material like a pair of chisels. A Wood drill doesn't have the twist, but it still has the shape of chisels to cut into the material as it spins.

By reversing the design of a twist drill, we create a type of rotary cutting tool.

When drilling into metal and a some other materials, the part coming out of the hole often looks like a Spring, but it may not have the appropriate tension. In order to temper the outside of a coil of soft steel, hardening it to a springy tension, heat it uniformly to a light red color and dip it in oil. Cotton seed oil works well for this. Reheat it, burning of the oil, and re-dip it, up to 3 times. This will leave the steel a light blue color. Large springs are often further hardened by heating to a clear red and plunging into boiling water.

Springs will stretch when suspending a weight and return to

normal when the weight is removed. The greater the weight the more they stretch. This allows us to make a scale for measuring weight by measuring how far the spring stretches.

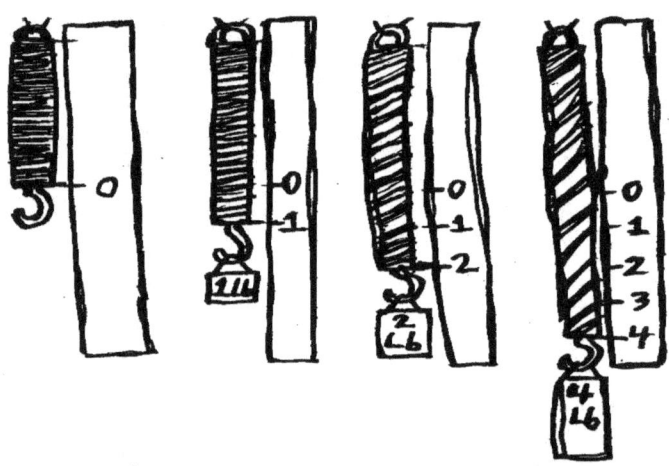

Springs can also be compressed at a similarly predictable rate.

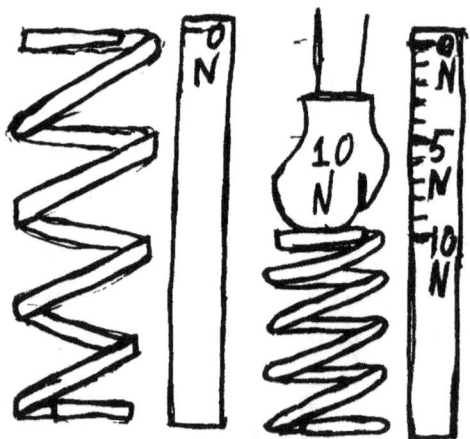

Using a wedge to redirect the force and a couple of springs as a   counter-force we can use a lever arm as a gauge indicator. Of course we could also have the lever moving some gears to turn a gauge dial at a different angle.

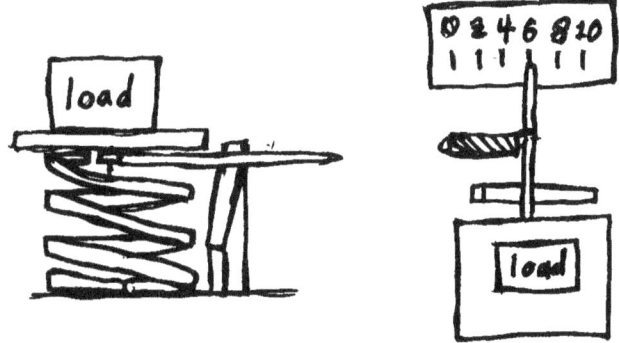

Springs can also respond to forces that tighten or widen their coils, and can be arranged in a spiral to take advantage of this. Clock springs and those used for "Clockwork" devices are a good example of this. Before the electronics age, there were a number of automatons [literally "self moving" devices or robots] that used tightened springs to turn gears and pulleys and move in response to various triggers. Of course with enough tension a spring can break or get a sharp bend, requiring it to be replaced or remade.

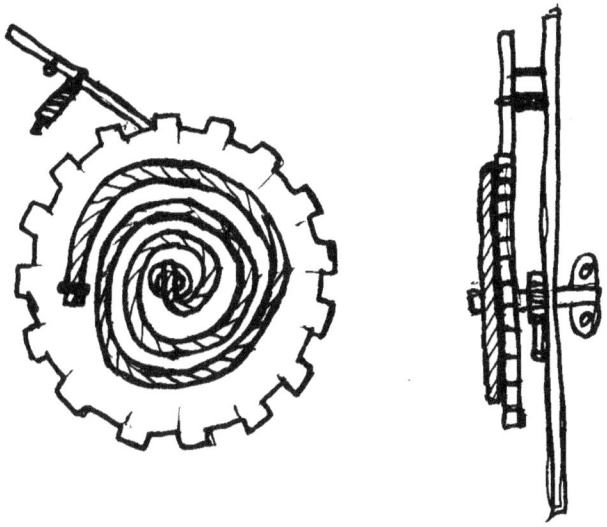

Putting tension on a spring and releasing it will cause it to vibrate at a Harmonic frequency based on the size of the spring and it's inherent elasticity & tension.

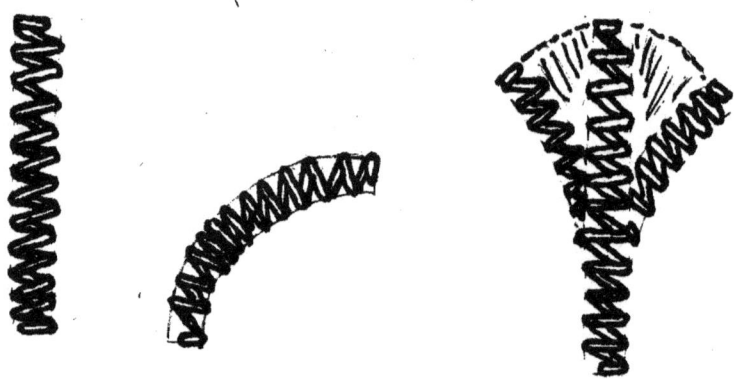

If you were to attach a pencil or pen to the free end of a vibrating spring and drag a piece of graph paper by it fast enough, you can get a chart like this one.

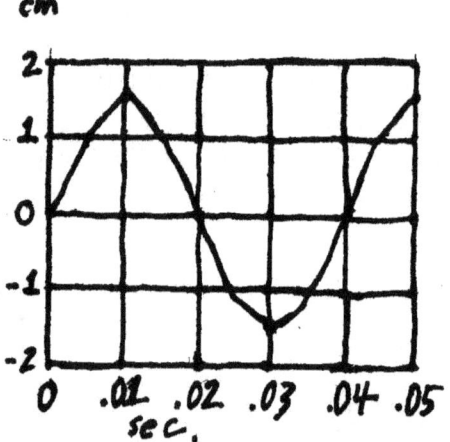

The distance the spring moves above the neutral position is called the Displacement or Peak amplitude. The distance between the highest point and the lowest point is called the Peak to peak amplitude. The Period [short for time period and represented by T] of a vibrating system is the time it takes to complete one cycle of vibration, moving back and forth one time.

The Frequency [f] is the number of vibrations per unit of time and one vibrational cycle per second is referred to as 1 Hertz [Hz]. f=1/T

Hooke's Law defines the interactions of a spring's vibration or simple harmonic motion:

•That inherent property of the spring can be referred to as the Spring constant [abbreviated k]. In order to vibrate there must be a Displacement force [x] and a Restoring force [F]. The restoring force should be equal and opposite to the product of the spring constant and the displacement force [F = -kx].

•The potential energy stored when that spring is displaced equals half the constant multiplied by the square of the displacement force [$\frac{1}{2}kx^2$].

•You can figure the exchange constant between potential and kinetic energy by taking into account the mass at the end of the spring [usually ignoring the mass [m] of the spring itself] $\frac{1}{2}mv^2 + \frac{1}{2}kx^2 = \frac{1}{2}kx^{o2}$ [x^o = peak amplitude].

•The speed [v] of that harmonic motion equal $v=\sqrt{(x^{o2}-x^2)k/m}$

•The acceleration is due to the force on the mass[F = ma] as well as that on the spring[F = -kx] for a hybrid formula [a = (-k/m)x].

Next is:

Chord & Wire:

For building and manipulating things you often will need
some kind of string; rope; or other chord.

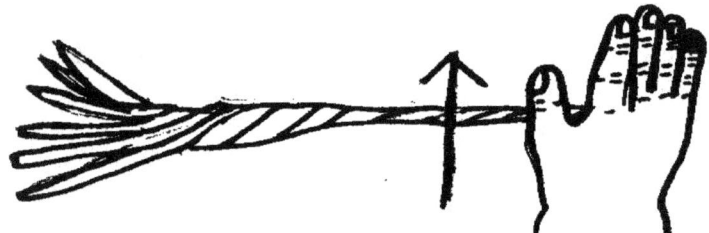

With soft fibers you can rub them against a surface and they
will be twisted and drawn out into a line. Yarn and many
other types of string are made this way, along with some
types of rope. Animal hair like wool sheered from sheep,
mohair trimmed from goats, and even angora combed or
brushed from rabbits or cats is commonly used to make
yarn. Silk from cocoons, or recently even from spiders, is
used to make a strong thread. The down from cotton seeds,
dandelion seeds, cottonwood seeds, and even from the
seeds of the cattail bulrush plant may be spun in this way.
For making ling spools of thread or yarn, it would help to
have a Spinning wheel or similar device.

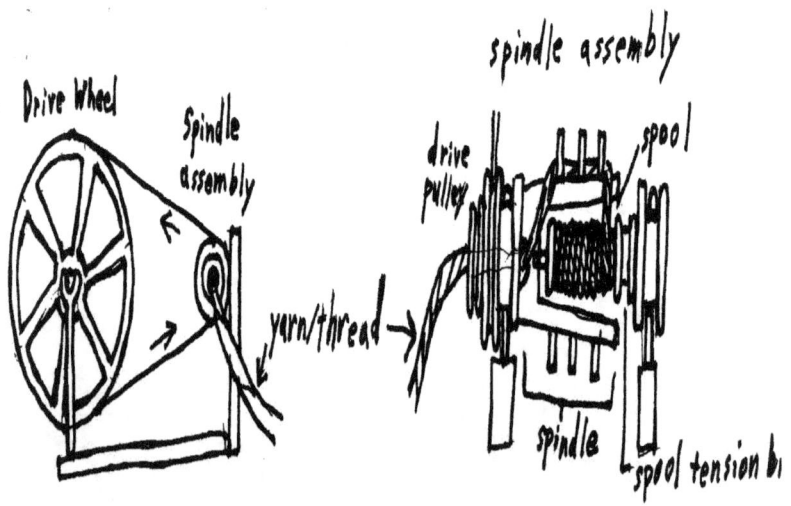

Fibers that are thicker and stiffer can easily be braided into twine using three at a time.

You can add in a new strand before a short one runs out by holding them together and braiding it in.

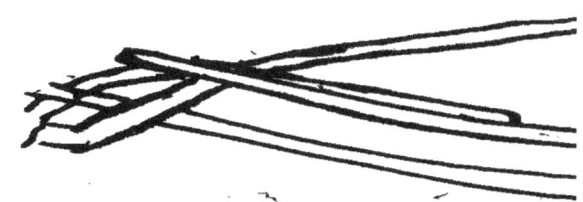

With practice you can develop the coordination for a four strand braid.

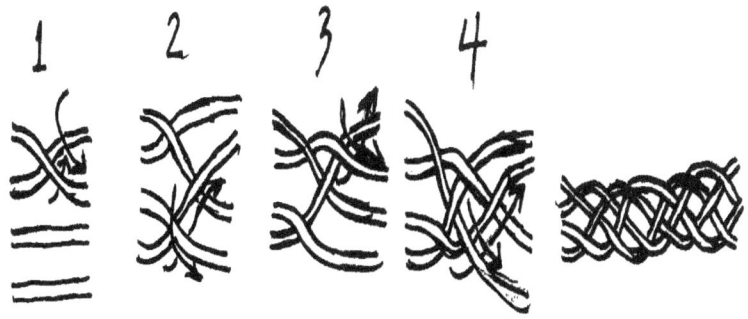

You could twist a few lengths of twine together to make a rope using the method from before, or you could put together a machine system to divide up the workload.

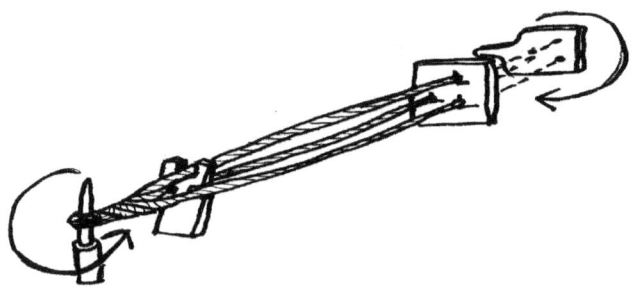

With a length of chord or loosely attached lever you can make a simple tool called a Pendulum.

A pendulum can be used with a graduated frame to make an easy to read indicator of how level or balanced a structure is in relation to gravity. Used in this way, it's often referred to as a Plumb-bob.

With a loosely attached lever and an adjustable weight secured to a frame by springs you have a metronome. A Metronome is used to maintain a steady rhythm for certain activities; including making music.

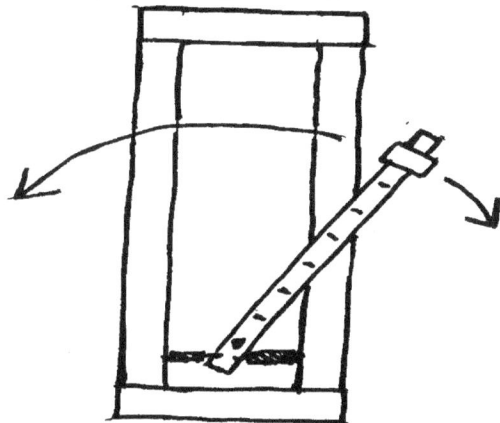

This principle has been used in clock-making to keep accurate time for centuries.

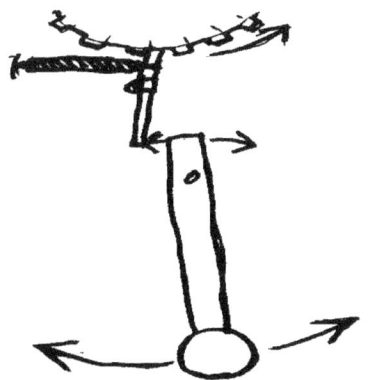

The harmonic time period [T] of pendulums can be calculated by considering the movement as a partial circle and comparing the length[L] from the connection to the weight and the accelerating force[g] of gravity. $T = 2\prod\sqrt{L/g}$

Once you have some cordage you could use it to make a net or some cloth. One of the oldest methods for doing so is by Weaving. There have been various designs for Looms to assist in weaving cloth developed by different cultures around the world. The basic idea is that several strands of chord or thread are attached to a support and every other vertical thread [the warp] is pull away in alternations so that a horizontal thread [the weave] can be passed between. Picking up every third thread or some other pattern can be used for different visual effects.

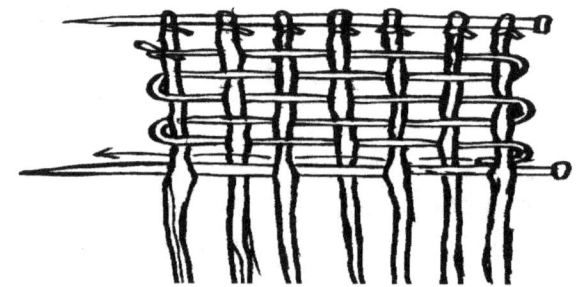

An alternate method called Macramé uses knot tying to link the threads in a network. Macramé is often used for making belts and artistic wall hangings.

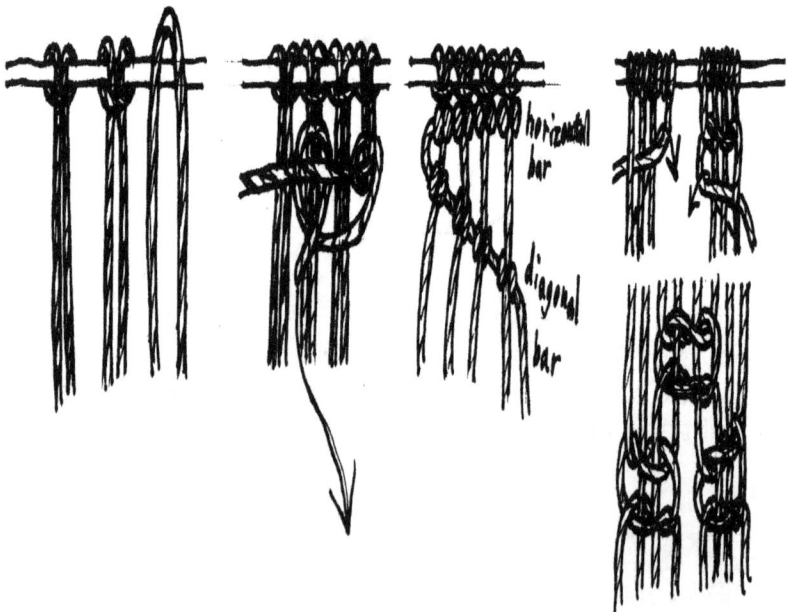

Knitting is a technique using a couple of rod with points [called knitting needles] and a single strand of thread; yarn; or other chord. You can   create a fine or loosely knit length of material. To prepare, you secure one end of the chord to one of the knitting needles with the loop of a slip knot. Then you make loose half hitch loops to the other end of the needle. Insert the point of the other needle into the first loop, and catching the loose chord with the tip, pull it through to form a new loop. Keeping the new loop on the second needle, insert the tip through the next loop on the first needle. It can help to pull the old loops off the first needle when you're done with them. Once you get to the other side and the first needle has been removed, you can insert it into the first loop of the second needle and work your way back to the other side.

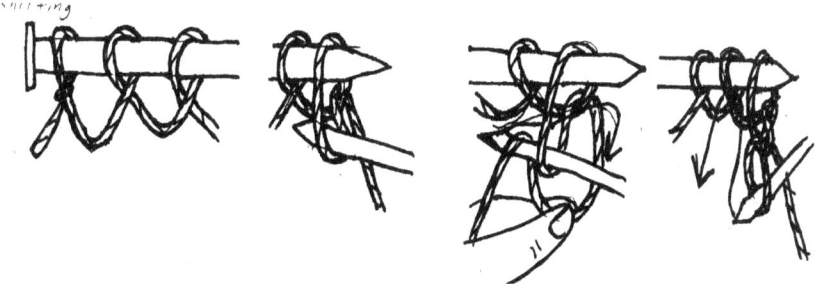

Crochet is a technique similar to knitting, using a single rod with a hook on the end [called a crochet hook or crochet needle] to create a chain of slip knots. To start the next row, insert the hook through one of the old loops and pull the loose chord through both loops. Making additional new loops in between connections creates a looser mesh and makes the work easier.

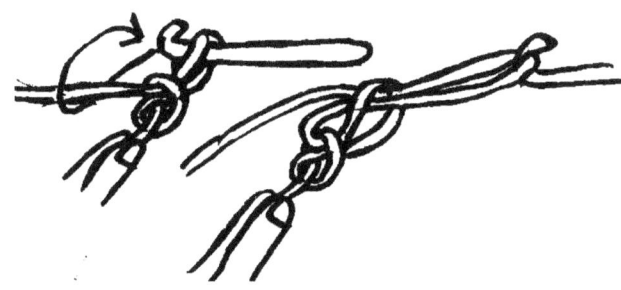

Different types of needles can be made of various materials, from metal and wire to wood and bone. Due to a tendency for some materials to crumble, you may need to use a notch instead of a narrow eye to hold thread.

Bending a segment of wire into a near circle, you can make "Jump rings" or you can continue the loop into a small coil or "Toroid".

Linking these together you can make chains. Soldering or welding the gaps in the rings/toroids can improve the strength of your chain.

By linking chains together you make chain-mail which has been used as one of the most basic types of armor to resist cutting. Many different styles have been developed over the years which reduce the size of the gaps between the links.

Paper is similar to cloth and chord in that it's usually made of plant fibers. Boiling wood shavings and other plant fibers into soggy pulp, it can then be spread over a cloth or mesh screen to be strained or pressed before drying in sheets. The old fashioned methods used in Asia tend to make paper that is either softer or stronger than similar weight & thickness made by faster modern methods.

Ink and pigments for writing, drawing, painting, & dying have been made from various materials from crushed stones or charcoal to simple plant juices. Pens and brushes have been made from sticks, wires, feathers, hair, and various other materials.

Glues and adhesives can be made from a number of materials as well. One of the most common is to mix bleached wheat flour and water. The sticky protein works well, but it stays water soluble. Powdered white or sticky rice mixed with boiling water works better and even when dry resists cold water. The sap from many types of trees is very sticky and liquid when heated, but when cooled becomes quite hard and repels water.

6 - Basic Construction Techniques:

Working with Grass, Trees, & Wood:

You'll need some cutting tools to do this. A hatchet or an axe will prove very handy. A saw or a machete may be useful as well.

The simplest type of structure can be made by leaning branches against the side of a cliff or better yet a partial cave under a rock out cropping. If the leaves or needles are angled down it will help to channel rainwater to the outside.

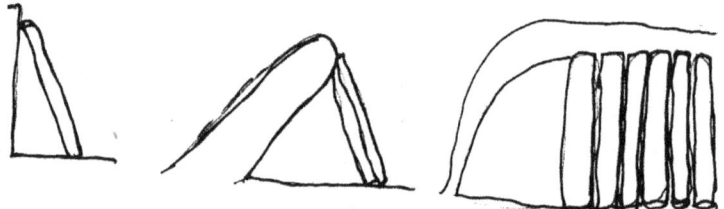

It's a bit wasteful but, the next easiest would be to cut a well shaped pine or other conifer of sufficient size where part of it is still hanging on the base. Removing some of the underside branches to make room.

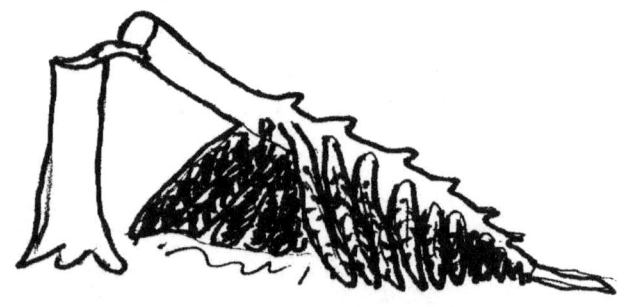

A similar shape can be made using a couple of forked limbs to support a much longer log. A few more supports will make it easier to stack leafy branches against it; or other covering materials.

If you want a structure made with logs to last for a long time, it's a good idea to remove the bark. The bark holds in moisture and provides a hiding place for wood eating bugs as long as it's attached to the log. For some trees the bark is loose & flexible enough to remove in large pieces or strips that can be used as shingles and outer wall coverings. A sharp machete or carving knife can be used for this or a draw-knife is designed especially for this purpose.

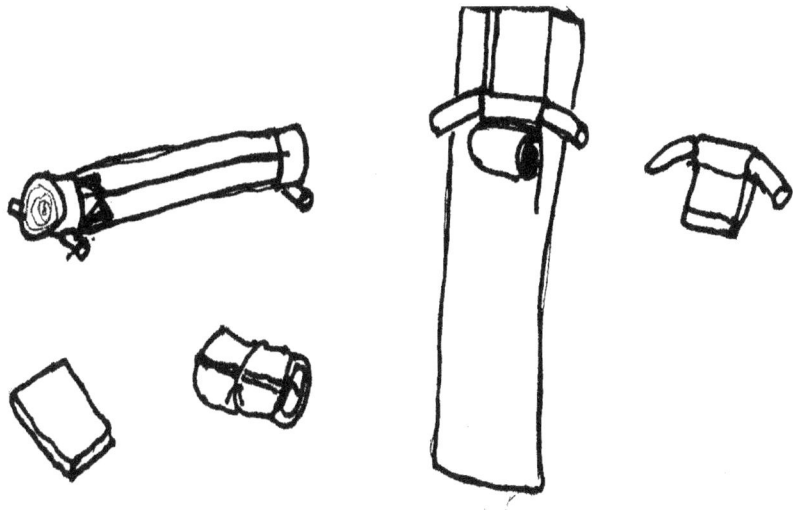

To make more complex shelters it helps to know how to lash the logs together.

Depending on the angle of the logs and their use, there are a number of different lashing techniques that can be used. For all of them you first secure the chord to one of the logs.

For Diagonal lashing you wrap the chord around the center first one way, than another to secure the logs, before wrapping it between them to tighten the chord and lock them in place. The turns of chord that secure things together are call Lashing turns, and those that tighten things up are called Frapping turns.

Square lashing, for logs set at nearly 90°, goes over one log and under the other.

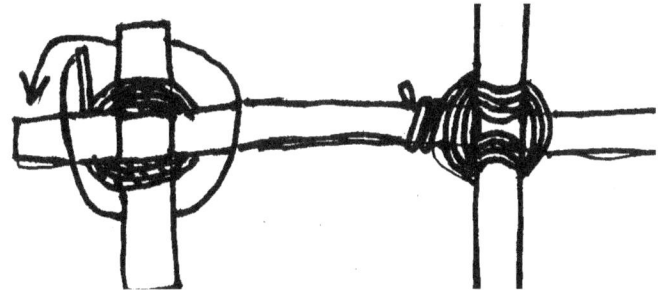

Shear lashing creates a flexible joint between two logs. Tripod lashing creates two flexible joints between three short or long logs.

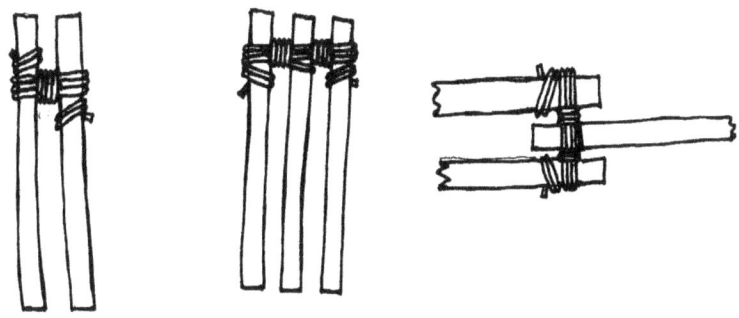

Just like the name sounds, you can use this to set up a tripod to hold a pot over a fire or to use as a lever fulcrum or other kind of support. With some leather or cloth to use as an outer cover you can just as easily set up a teepee type structure. Leaning a few more poles against that support, you can make the teepee much roomier.

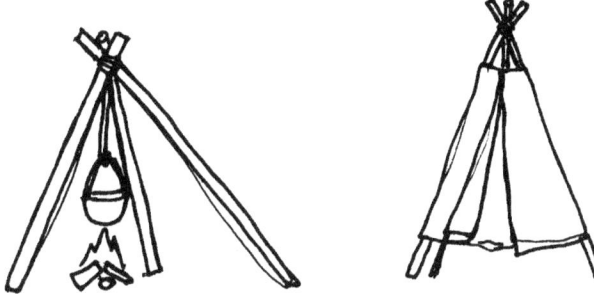

With a couple of logs made from forked branches set into the ground as supports, and at least three other poles, you can make the frame of a lean-to. Covering the backs and sides to keep out the wind and rain and placing a few lean-to structures on opposing sides of a camp fire can make for a cozy situation.

By doubling the lean-to we get a basic tent shape. If you use four support logs and four crossbeams, leaning branches in from three sides, it takes the shape of a simple lodge.

With longer poles and more supports, you can expand on this theme and create an actual liveable space. Just remember to leave an opening at the top for ventilation if you're going to have a fire inside.

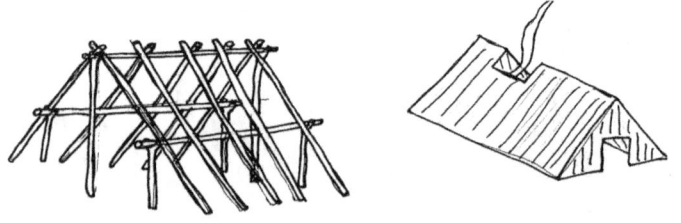

These structures can get quite elaborate.

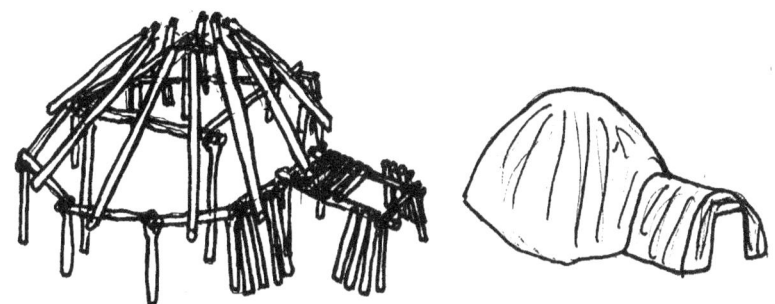

If flexible or curved poles are available, there are even more options for the frame of the building.

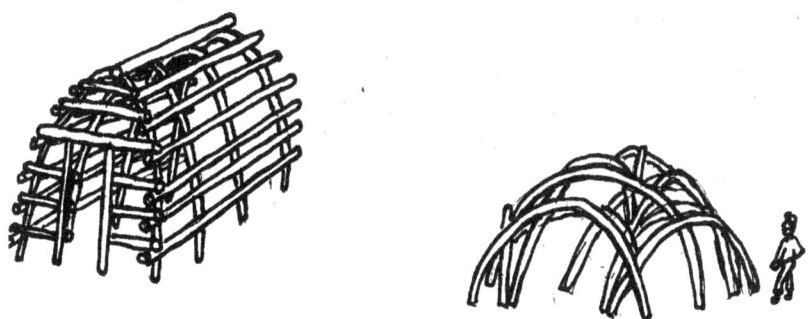

For various reasons it can be a good idea to put a structure on stilts and raise the floor off the ground. This also allows you to build over a body of water.

In swampy areas, stilts alone may still sink. Additional support can be arranged to spread out the weight, reducing the pressure on the ground much like snowshoes.

In a river, it may be necessary to secure the ends of those stilts in buckets, pots, or baskets filled with stones or other weights to prevent them from being washed away with the flow.

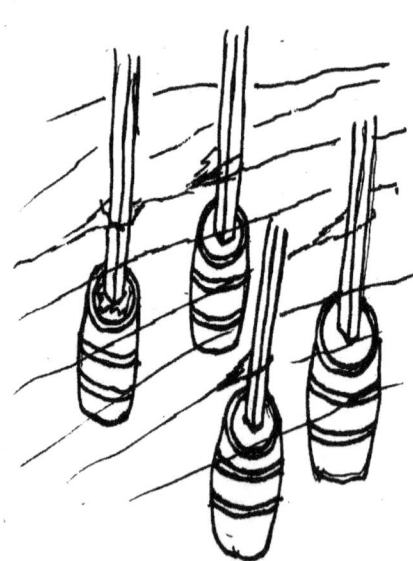

A common covering used all over the world is thatch. Thatch is basically the stems of grass or straw from grain that has been bundled together. If you want to keep rodents and birds out of it, a good idea would be to remove all the seeds and leaves.

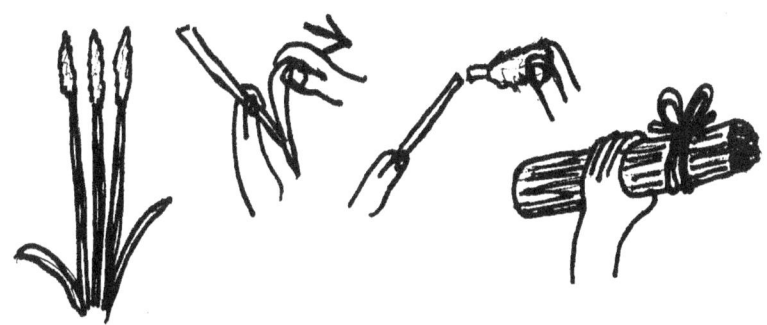

To use it as the covering for a structure, it's best to secure it in the middle & at the top and to overlap them so they can channel rainwater to the outside. This can also be done with pine boughs or those from other coniferous trees.

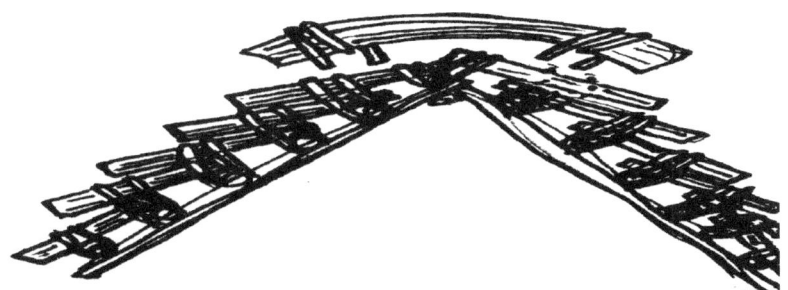

A bed made from such soft materials that have been overlapped can be very comfortable.

It's possible to make some easily assembled prefabricated structures using beaver-mats. Beaver-mat refers to a layer of overlapped thatch or pine boughs or other material that is sandwiched between two log/pole frames.

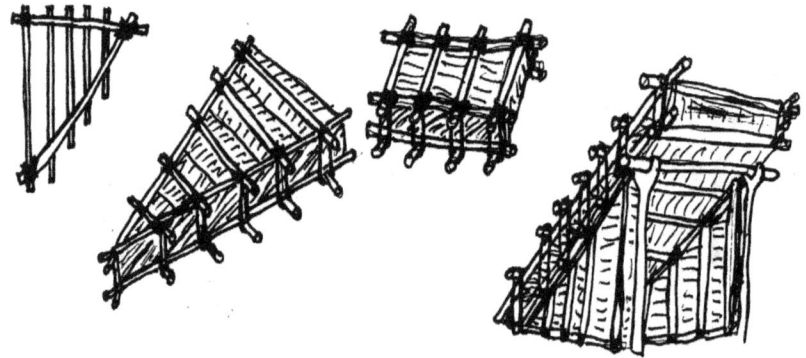

A slightly more complex design is the stick shack. Against a frame of poles that are set into the ground, bundles of sticks [technically they're called "Fagots"] are laid like bricks. Afterward, any type of roof covering can be used. It's a good idea to lash a cross pole near the top to keep the frame from spreading out over time, especially if it's going to be a long-term project. If you want to seal out the draft, you can line the inside with clay or mud.

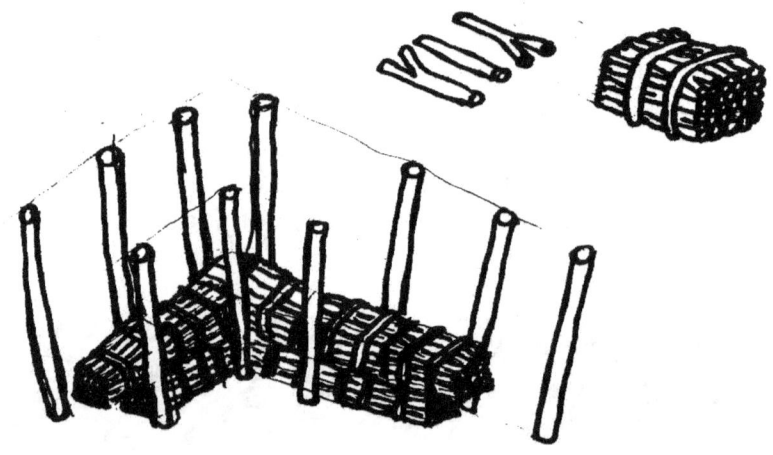

By cutting notches out of logs, they can be used as is for ladders. This also provides a more secure connection for attaching to other logs. By overlapping notches we can put logs together with very little space between. This is the key to making log cabins.

If you line the inside with mud or clay you can even make a fireplace out of logs.

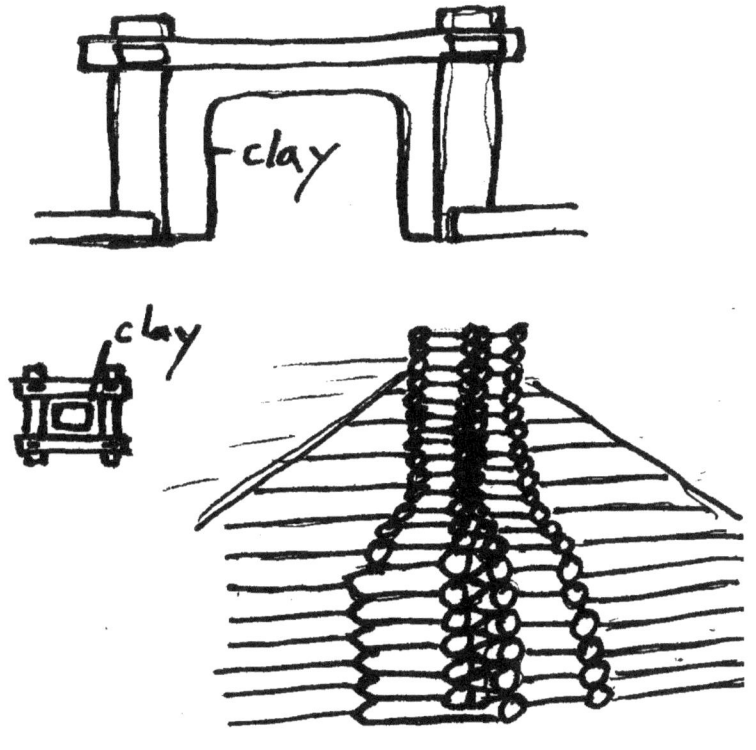

It's important to start the cut on the diagonal for lower resistance.

With a little more work, you can square off that side, flattening it.

This can be done with an axe, but it's a little easier with a tool called an adze.

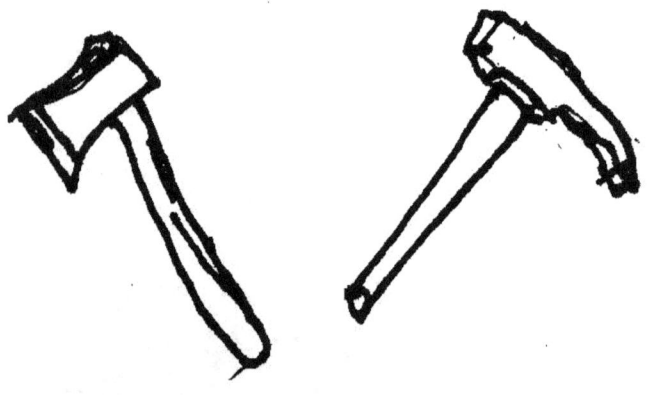

Large logs can also be split with a pair of axes. You can accomplish this by hammering in a number of wedges as well.

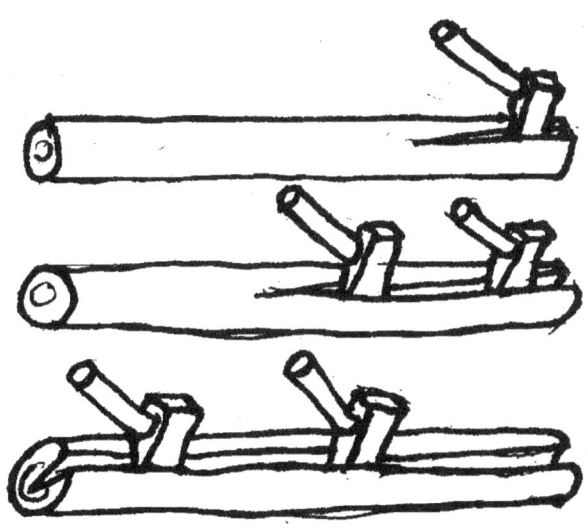

With some help and a big enough saw, giant logs can be cut into boards.

A single log of sufficient size can be cut into many boards of various thicknesses, widths, & lengths.

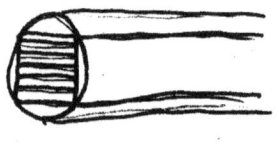  

If the logs in an area aren't thick enough, they can still be cut into uniform thicknesses and glued together. Turning the grain different ways for each layer to maximize strength makes plywood (ply- means layer).

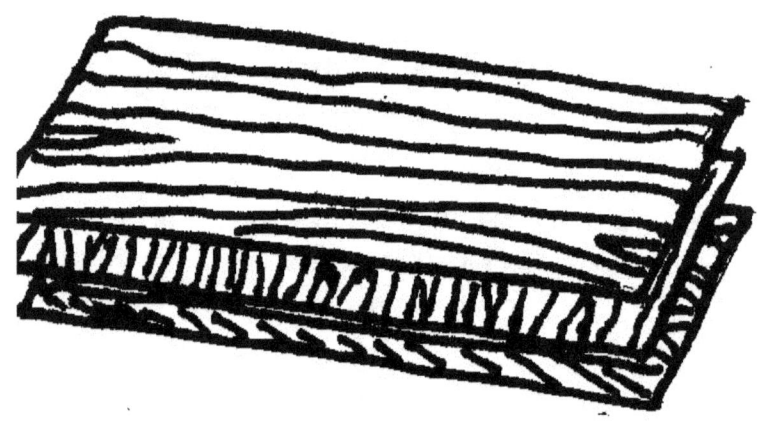

The easiest way to make three dimensional structures with boards is to connect them at overlapping 90° angles.

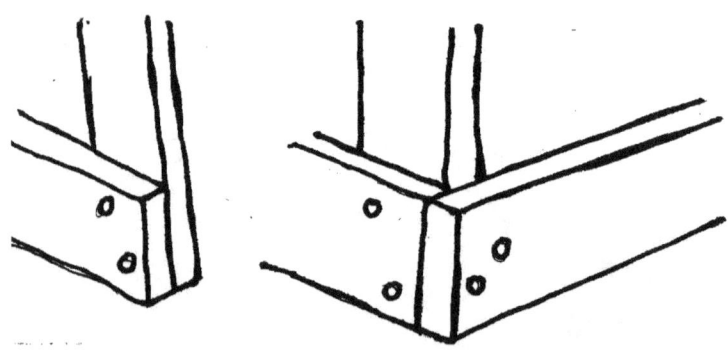

If that's not practical, it's possible to link them with a bracket or other connecting piece.

It's also possible to cut a notch so that one board may be supported by another.

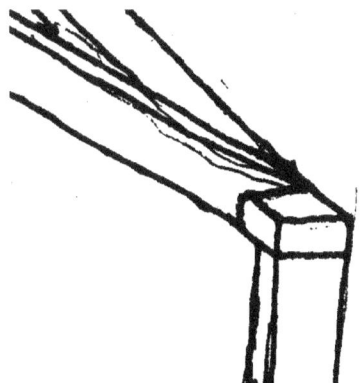

When it isn't practical to reach the other side of a board for installing a connector, there are a couple of options. One is to "toenail" the top board by putting a nail, screw, or other connector through the two boards at an acute angle. The other is to use a secondary block of material and more connectors.

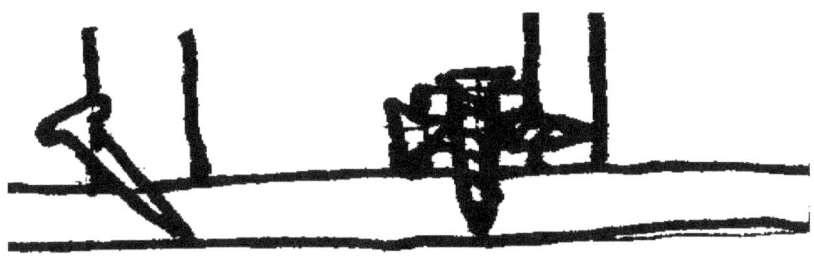

In order for structures made from boards to last a long time, it's important to protect them from wood eating bugs and moisture induced rot. One way to do this is to set a bottom support beam or "sill" on top of stone supports or wooden posts or piers. This is the basis of the pier & beam system.

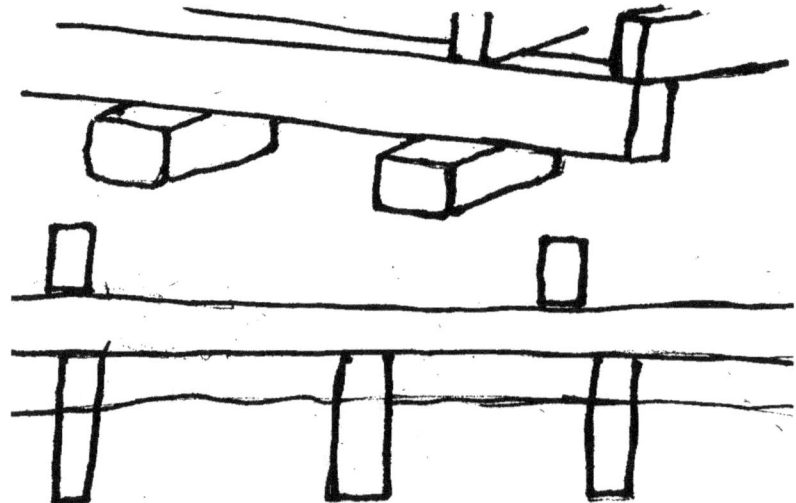

Another option is to build up mud hills to support the sills. If the bottom boards are treated with creosote, a solid or liquid condensed from wood smoke that is also called wood tar, they tend to resist rotting and repel wood eating insects.

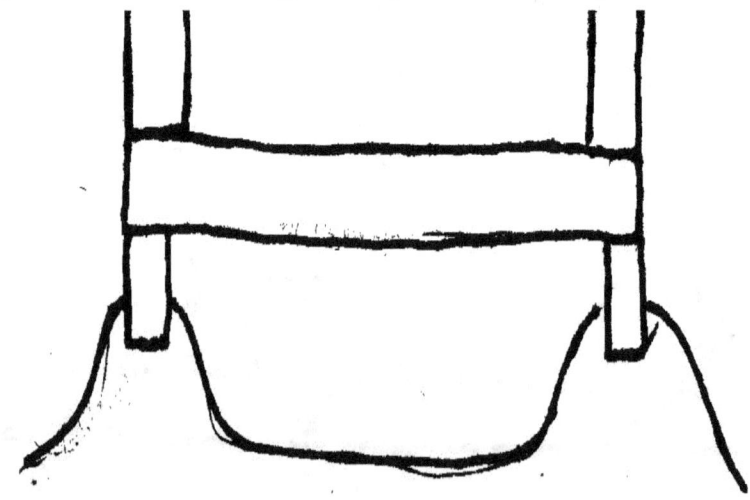

It's also interesting to build a structure in the trees. Here's one method:

Once you have a floor it's fairly easy to frame up the walls and a bit of a chore to put up the frame for the ceiling. Of course if you have help and scaffolding, or at least a pulley system and some ramps, it's easier.

Each board in a house frame has a name based on where it's located. The bottom boards are called the Sills. Those going across the sills are called floor beams or floor Joists. Those coming up at the corners are called Corner posts. Those coming up between the corner posts are called Studs. Between the studs and between the studs & corner post are boards called Ribs. The ones on top of the exterior walls are called End plates and Side plates. Any boards connecting the side plates in the middle are called Collars. The board at the peak that runs from end to end is called the Ridge. The boards that hold up the ridge are called Rafters, and those that run between the rafters are called Purlins.

Another strategy used by a number of cultures, is to put the roof up on supports first and then install a base floor and shape the space with walls last.

First, raise the center ridge on posts, followed by a couple of collars.

On the ends of those supporting collar boards mount the side plates.

Mount more collars between the side plates and end plates on the ends. From there you can connect the rafters between the side plates and the ridge.

Once the rafters are attached you can attach purlins on top of them, and roof covering material can be attached directly to the purlins. The covering can be simple thatch. The thatch can be covered with clay or mud to increase water resistance and to attach additional covering materials.

A covering commonly used for wooden shingles would be shakes. A Shake is a thin, wedge shaped piece of wood cut from a short section of a thick log. These can be cut with an ax or a saw, but there is a tool called a froe that is designed for this purpose. A Froe has a blade that is shaped like a hoe, but it's angled to be thrust or hammered into the log segment along the grain.

Another option is bamboo. Splitting a bamboo pole in half and remove the nodal membranes. Mount the bottom layer open side up and angled with the slant of the roof. Mount the next layer open side down to overlap the edges from the bottom layer. Another piece along the ridge finishes it.

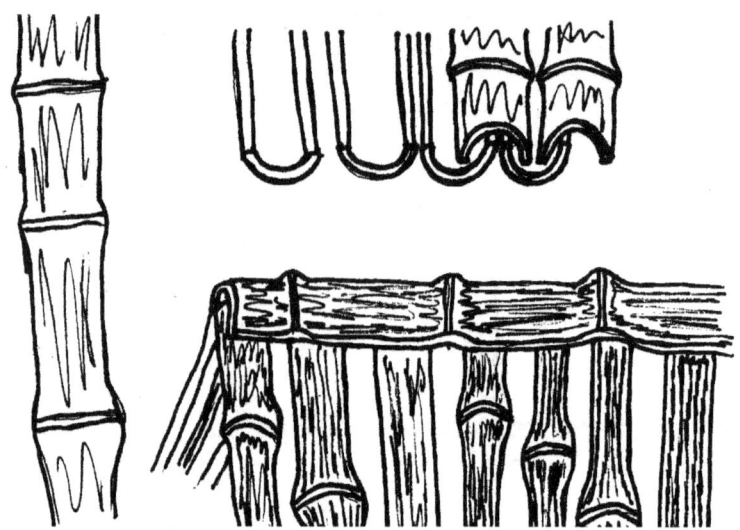

Another option for roof shapes is a gambrel roof that has two parts for each rafter. This creates a double slope which leaves more room for storage or open space near the ceiling. Some varieties of this design can be used without collar boards.

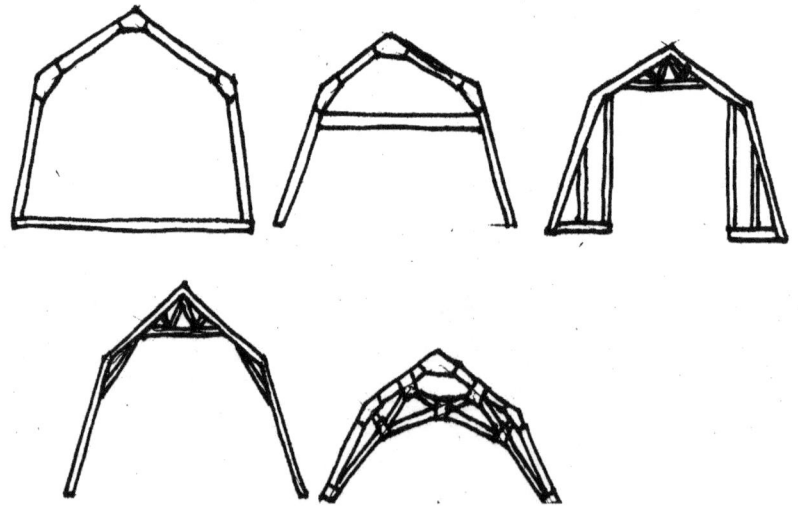

For thatch roofs, it's better to have a steep roof of at least 45° for rain water to run off & dry quickly. For shingled roofs, they need to rise up at least 1 foot for every 4 feet of of running to the side or about 14°. Clay roofs need to be almost flat to keep the clay from being washed away.

For a structure held up by support columns, the sills and some of the floor joists can be attached directly to those supports.

Of course with so much riding on them it's important that those supports be sturdy and well anchored in the ground, with up to 1/3 of their length below the surface.

There are many ways to attach to the frame of a structure and frame out walls with boards. One way is to chisel or cut out notches where the boards would overlap. If properly done so that they fit snugly together, rather than weakening the boards, they reinforce each other as though they were part of each other.

Another way is to drill and carve out a hole all the way through one board and shape the end of another one either to exactly fit it, or to push a little bit past, and to lock it in place with a peg.

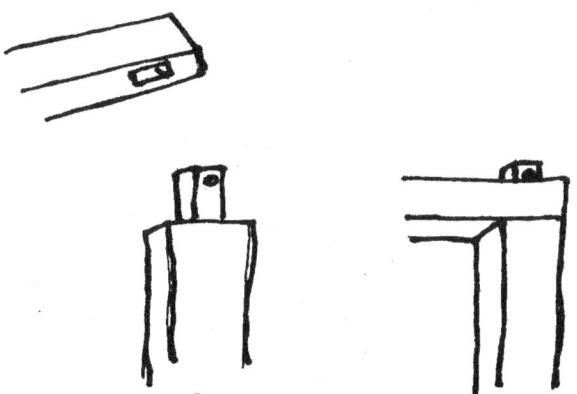

The space in between can be filled in with a variety of materials in addition to or in place of covering it with boards; panelling, or for walls that won't be exposed to weathering, even paper.

The easiest type of door to make is a sliding door.

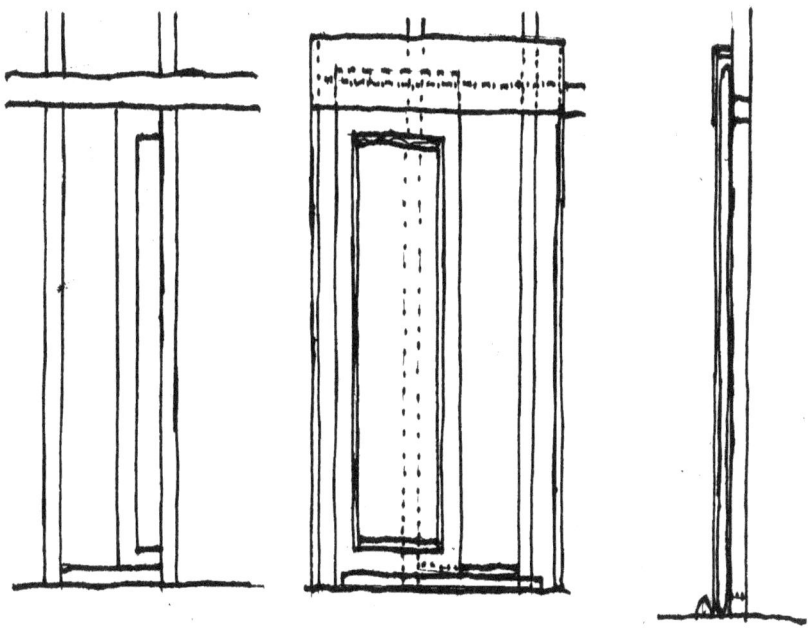

All that's necessary is a frame near the top to keep it from falling over and a partial track for it to slide in.

There are a number of options when it comes to mounting a door to swing on hinges. The easiest is to use a shear lash.

Next would be to attach the door to a rod or dowel and carve notches or drill holes into the door frame for it to fit into.

It's a bit more work to shape a peg and hole hinge for it to swing on.

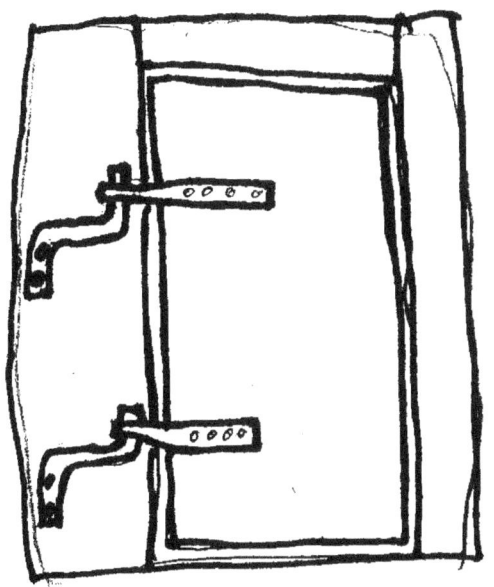

There are many ways to shape the doorway so that these swing to one side instead of the other and to minimize drafts. All vertically mounted sideways swinging doors tend to warp over time due to the pull of gravity and especially in damp or hot   conditions.

There are many ways to design latches for these doors.

The simplest is a lever with a hook and a peg for that hook to catch on.

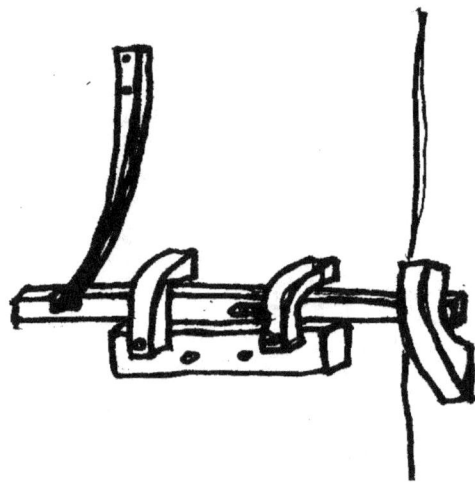

With a piece of wood that has been planed down enough to be flexible, springlike tension can be applied for self closing latches. Strings and counterweights can also be used in various combinations.

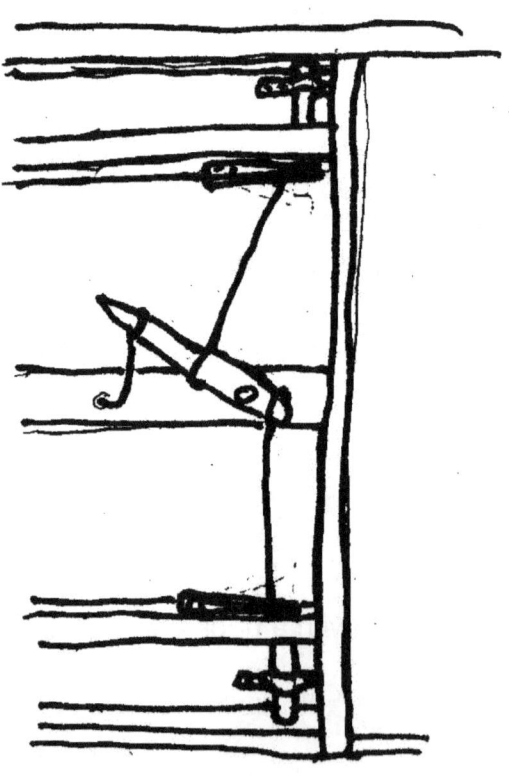

If a part needed to manipulate the latch can be removed or concealed, it's effectively locked to anyone who doesn't know the secret.

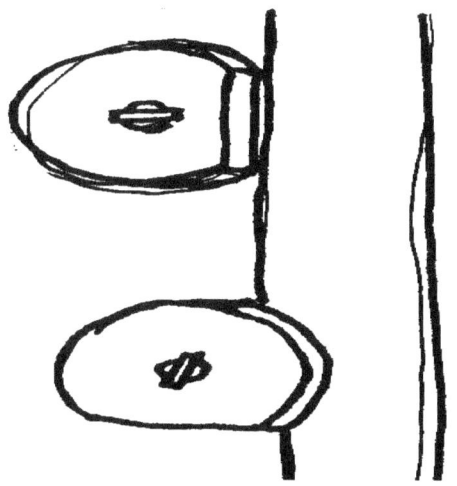

Working With Mud & Clay:

For areas with minimal rainfall, many of the buildings are made from mud bricks with a mud or plaster coating.

Hundreds of years ago many of these were dug out of lake or river beds that had nearly dried up for the year. Using sticks or boards to make a frame allows rectangular bricks of uniform shape and size to be turned out in large batches.

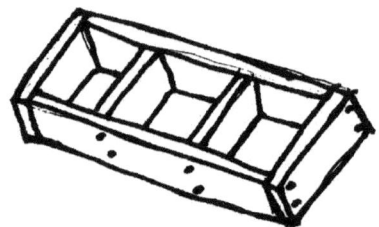

The mud is mixed from local soil and a small amount of water until it is very thick, but can still flow out of a bucket. Any grass, roots, leaves, or twigs that can be incorporated increase the durability of the finished product. The bricks are allowed to dry slowly before the frames are removed. If the can be fired afterwards, they'll be more weather resistant. A similar mud mixture can be used as a mortar to join the bricks and plaster over them, though the outer plaster may need to be changed every spring.

It's common to put doorways on the side at ground level.

On the other hand, in extremely arid areas and especially for buildings in the shelter or a cliff, some cultures used ladders and put their entrances on the roof.

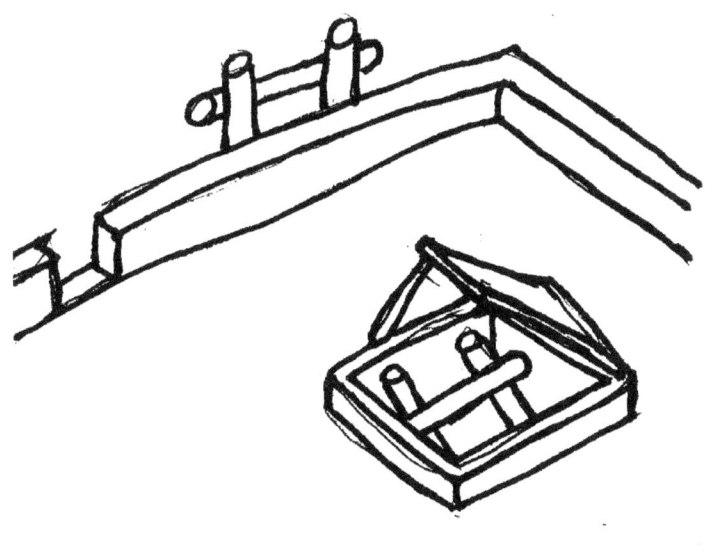

One type of mineral found in some soils and rocks is clay. Clay exhibits plasticity, a tendency to become malleable [easily shaped] when warmed and worked between the hands, after being mixed with water in certain proportions. When dry, clay becomes firm and when fired in a kiln, permanent physical and chemical changes occur in a process called vitrification [literally meaning "process of becoming glass"]. These reactions, among other changes, cause the clay to be converted into a ceramic material.

In order to make raw clay, dug from the ground, into a workable material:

> 1.First break it up into small pieces. If possible smash it to powder and filter it through a screen for better quality.

> 2.Mix it with water until it's a fairly thin liquid and let it set and settle for 8-12 hours.

> 3.Pour off the clear water and feel by hand for any hard pieces of rock, root, twig, or leaf. Remove any such pieces and pour or shovel the soft clay into a cloth bag and close the opening. Place the clay filled cloth bags into a bucket with a hole in it and place a weight on top of them.

> 4.Open the bags after 12-24 hours and take out the workable clay.

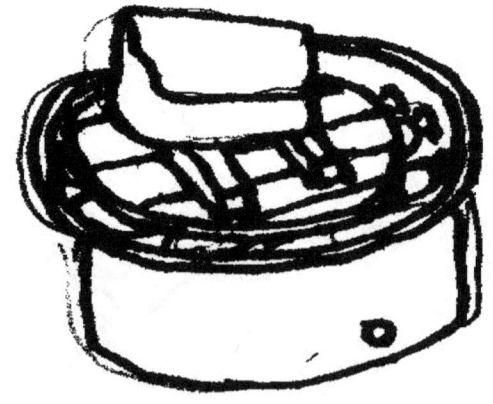

This clay can be easily shaped by hand or with various tools. Two pieces that have partially dried can be joined together by working the edges and applying slip. Slip is a liquid mixture of water and clay. Once pieces have completely dried, they can still be carved with a knife or drilled. After that they are ready to be fired in a kiln.

There are various designs for kilns. The important thing is to heat the clay/ceramic items as evenly as possible and avoid drafts.

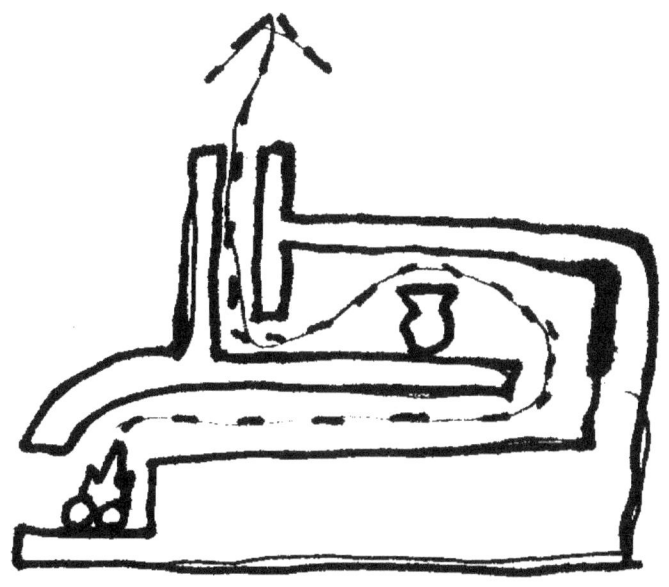

After an initial firing of 12-24 hours depending on the size of the items to be fired, all of the moisture should have been driven out. At this point they can be painted and glazed.

Many paints and pigments will change colors during the second firing. Glaze in this case refers to a material that can be painted on the fired clay and will melt in the kiln, forming a smooth and watertight seal over the surface as it cools. One of the most convenient glazing materials can be made by mixing wood ash with water. The second firing should take 2-4 days.

If a piece gets a small crack during the first firing, it can be fixed. Smash and grind another piece into powder and mix it with a glue made from powdered rice and   boiling water to form a paste. Fill the crack with this paste before painting and   glazing the pottery, and after the second firing it should still be able to hold water without leaking.

In the event a proper kiln is unavailable, by smoothing the surfaces and firing pottery in a bonfire or fire-pit, it can still be partially vitrified and made fairly watertight. First, rub or scrape the surface with a hard tool to smooth or "Burnish" it.

burnishing

The next key is filling and supporting those pieces with dried manure or similar material which will burn in the fire and concentrate that heat against the inner and bottom surfaces of each piece.

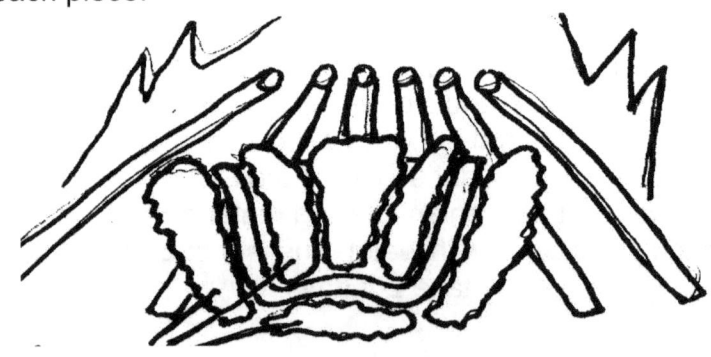

Shaped or moulded into rectangular blocks, clay is fired into modern bricks.

In most industrial kilns there are a few places of temperature extremes where clay pieces often crack. In large pottery companies, these waste pieces used to build up to the point they had to hire someone to smash them down in order to save space. Centuries ago, some one came up with the idea to recycle this material by grinding it into a powder and mixing it with a little water and fresh clay. The resulting material was a bit coarse and grainy, but the pottery made from this clay was able to survive in the hottest parts of the kiln without cracking. This "Fire clay" or pre-fired clay is now used to make "Fire brick" for fireplaces, kilns, and other high temperature applications. Fire clay to be used as mortar between fire bricks and for making high temperature linings is usually sold as an unmixed powder.

Spread into square and rectangular sheets clay become tiles to be used as floor, wall, and roof coverings.

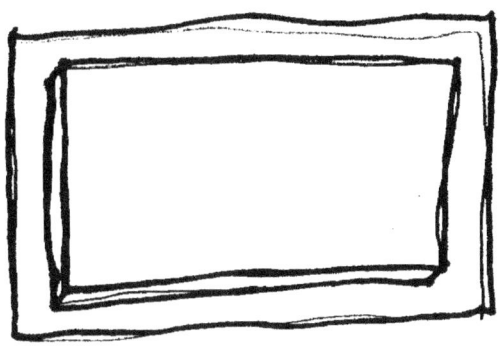

Using the curve of bamboo as a mould, they become the wavy roof tiles used in various countries throughout the world.

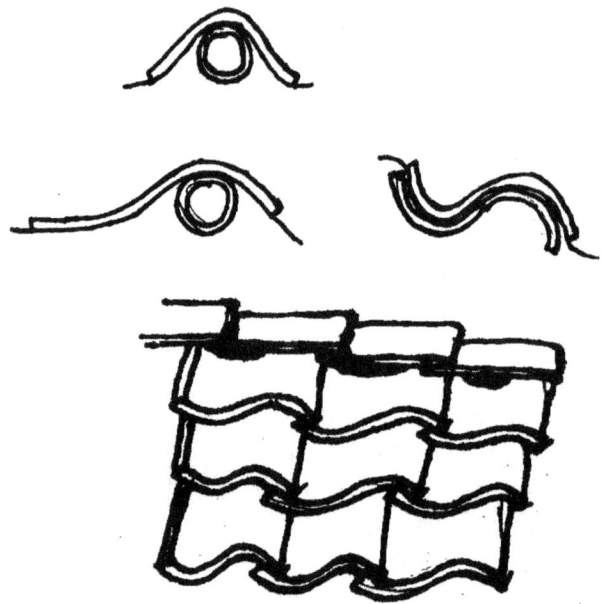

Special variations are made for the end tiles at the bottom edge of a roof.

Stone Work:

It's a bit labor intensive to shape and move it, but stone is one of the most common and enduring building materials available. For stability, rock walls should be made at least two stones thick; with the taller sides to the outside so that each layer of stone is leaning toward the middle of the wall.

Be careful to avoid lining up the joints between the stone to form vertical cracks that weaken the wall. Align each stone so the weight of the upper stones holds it in the wall instead of pushing it out.

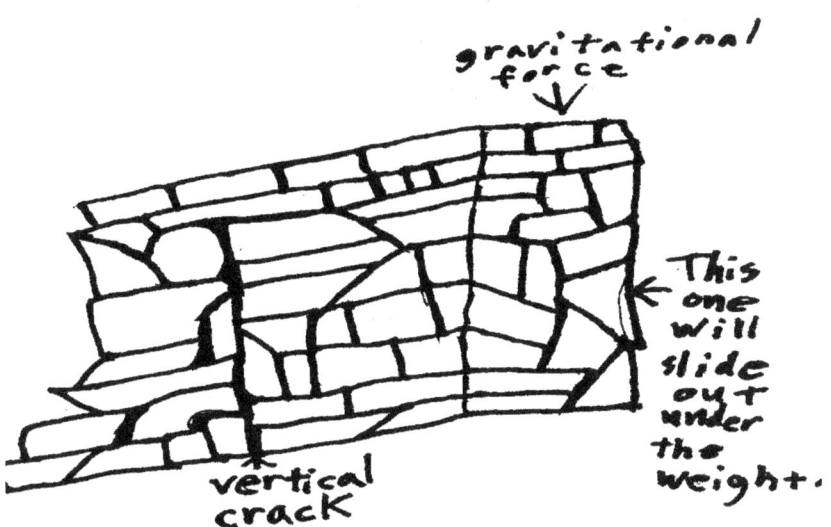

As temperatures rise and fall, causing materials to expand and contract, the stones in an un-mortared wall are always shifting back and forth. The freezing and thawing of water under the surface of the ground can thrust up under parts of a wall several inches. This same phenomenon causes many of the potholes in paved roads all over the world. In order to keep this from messing up your wall, it helps to dig under the frost-line (the depth to which the local soil will freeze) and put your largest and heaviest stones on the bottom.

12"
30cm
foundation
wall

|— 24" —|
60cm
dry stone
foundation

Any place where a wall turns a corner or has another wall intersecting it, its strength and stability is increased. For a long narrow stone wall it's a good idea to include at least one buttress in the design. A "Buttress" is a smaller and shorter support wall. If a wall that is one foot wide and eight feet tall is going to extend 24 feet (~7 meters) before it turns a corner it's going to need a buttress or intersecting wall near the center (about every 12 feet or ~3 ½ meters) that reaches within 2 feet of the top.

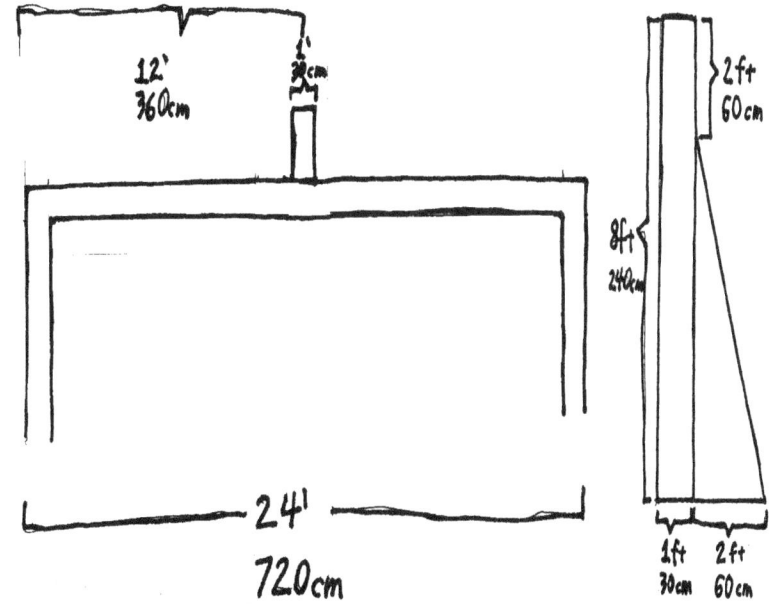

Otherwise you'll need to mortar the stones or make the wall thicker for support. It's important for the stones at the corners and intersections to overlap with clean lines as much as can be accomplished.

There have been various things used as mortar for holding stonework together over the centuries. Sand, powdered lime/limestone, and clay or Portland cement have been the most common. They can be used separately, but combined they form a stronger bond. Interestingly enough, Portland cement itself is often made from calcined lime mixed with clay and other ingredients in different ratios. Here's a blend that you may find useful.

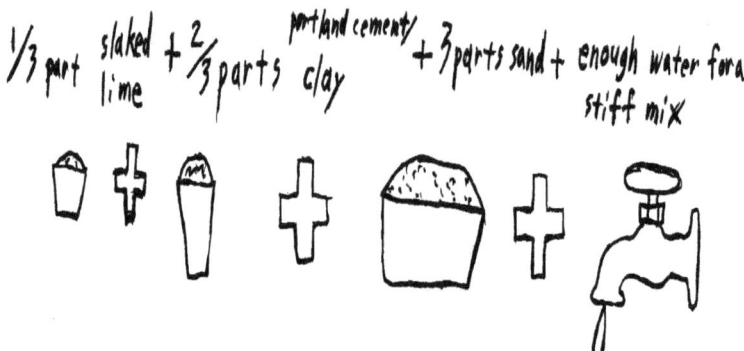

For aesthetic purposes, many people find it important to "strike-up" the work. That means to remove the excess mortar that bulges out between the stones while it's still soft (called the "green" stage) by brushing it out with a stiff bristled or wire brush.

The least complicated way to cover a doorway or walkway is with a lintel stone. A "Lintel" is a horizontal block that spans the space or opening between two vertical supports.

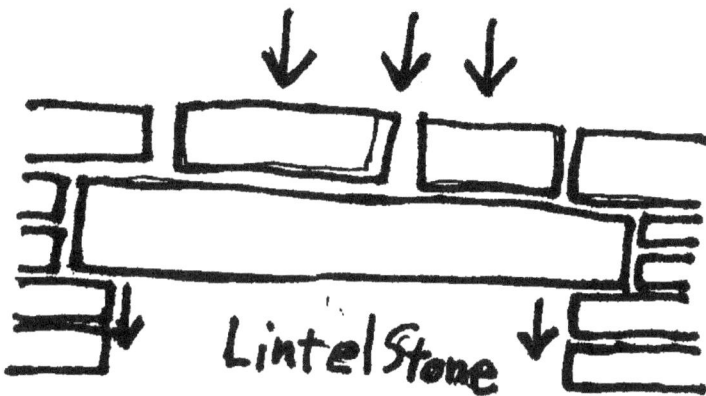

Unfortunately lintels sometimes break and fall along with any unsupported stones. Stones that continue sticking out, held in place by counter-leverage from the stones above them form Cantilevers.

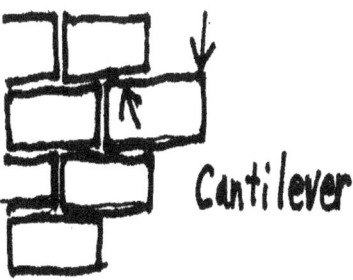

Stacking the stones in cantilever fashion from both sides you can make a Cantilever arch.

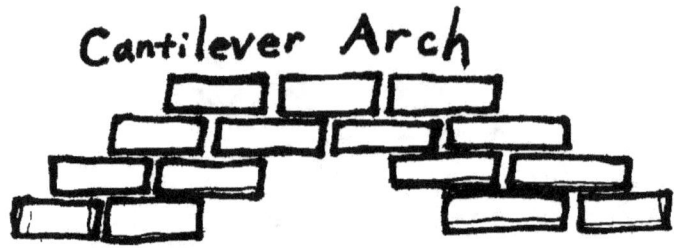

An easier way would be to lean two appropriately shaped stones against each other in an inverted "V". This angle redirects the downward force toward the center and guides the arrangement of the other stones into cantilevers.

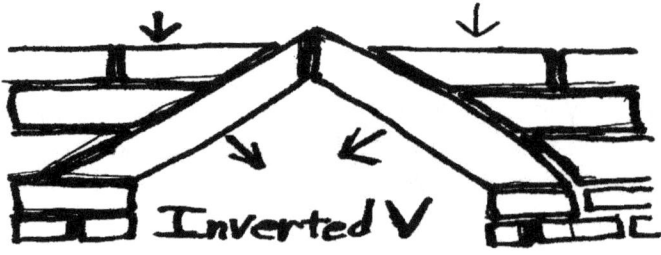

To make more complicated structures you'll need a specially shaped stone called a keystone. Basically a "Keystone" is shaped as a trapezoid or a trapezium.

With one or more of these you can easily build a Semicircular arch. The angles of the keystone turns downward force into sideways force and angles sideways forces downward.

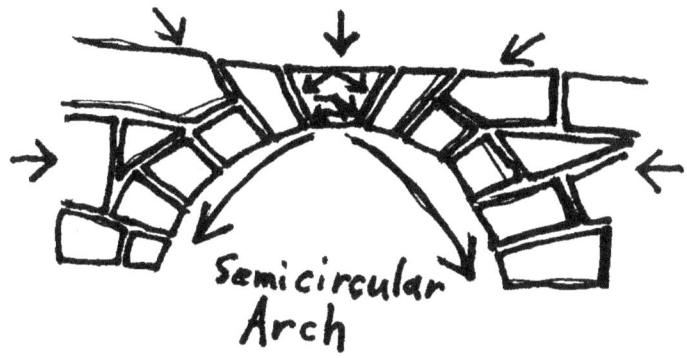

Semicircular
Arch

It helps to use boards and poles as temporary supports.

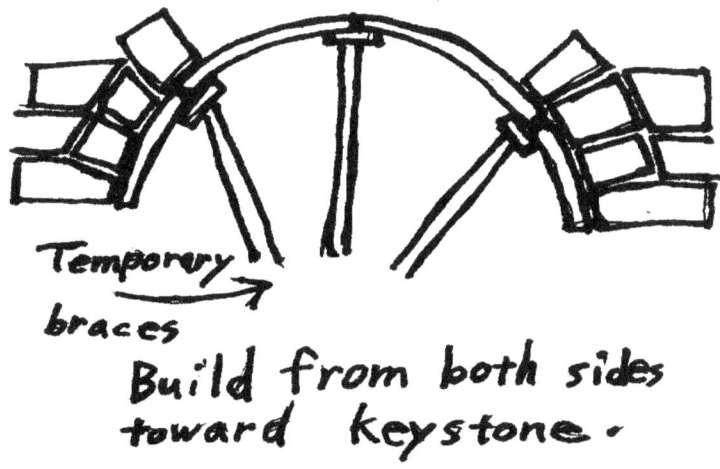

Temporary
braces
Build from both sides
toward keystone.

With just one and carefully planning the layout of your other stones you can make an elliptical or even a flat arch.

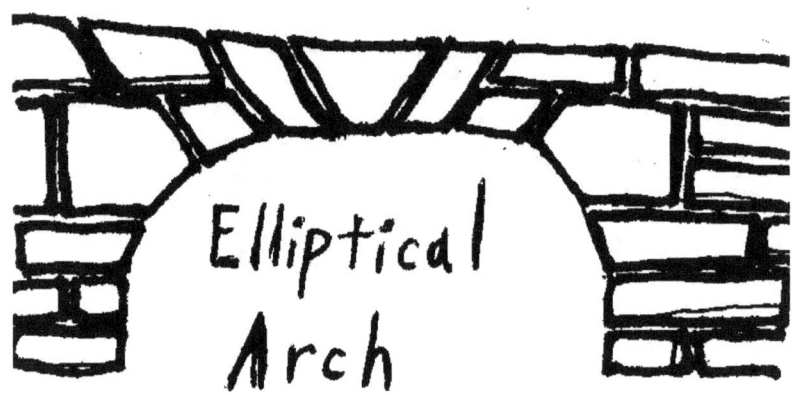

Continuing the arch into a circle you can easily line a hand dug well, keeping dirt and mud from pushing in and clouding up your well water.

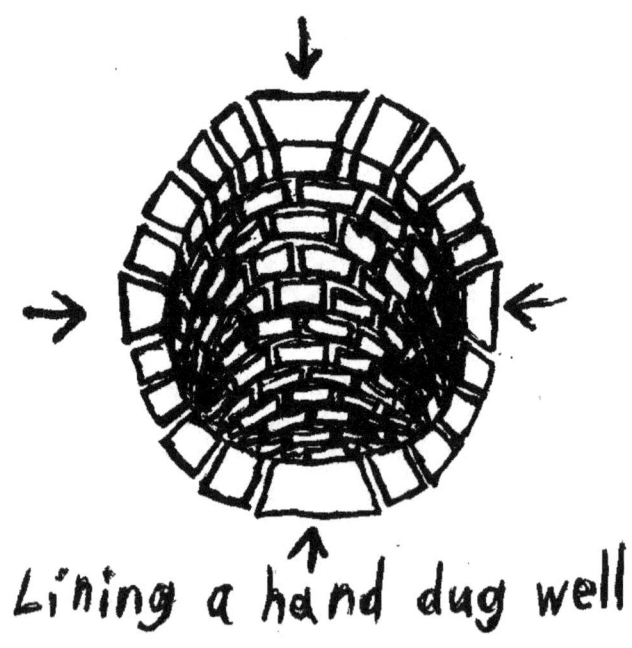

Over time terrain features can change, especially with rain softening the ground, even hills and cliffs can collapse. A properly built retaining wall can help minimize or prevent this.

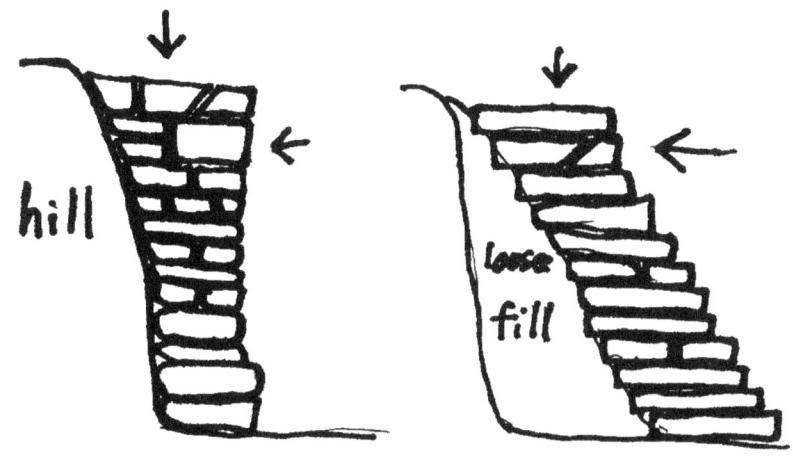

It's important to have the wall lean into the hill it's supporting. If loose material like gravel or sand is used as landscaping filler, the wall needs to lean in even more.

If the filler is going beside an existing hill, it's important to dig steps into the side of the hill.

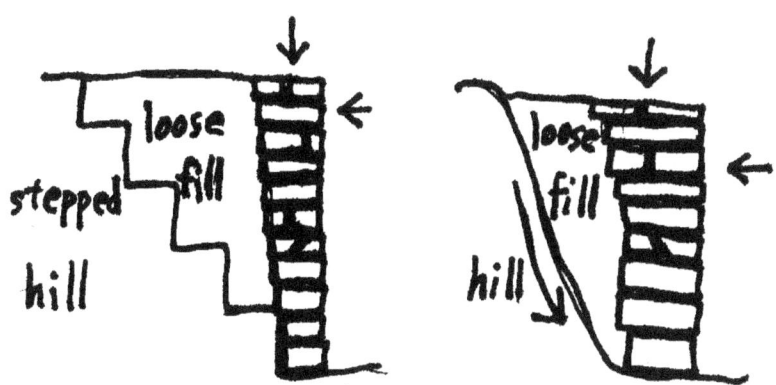

Without those steps to hold it, the filler material will be pressing against the bottom of your retaining wall like a wedge.

In order to shape those stones a number of tool can be useful.

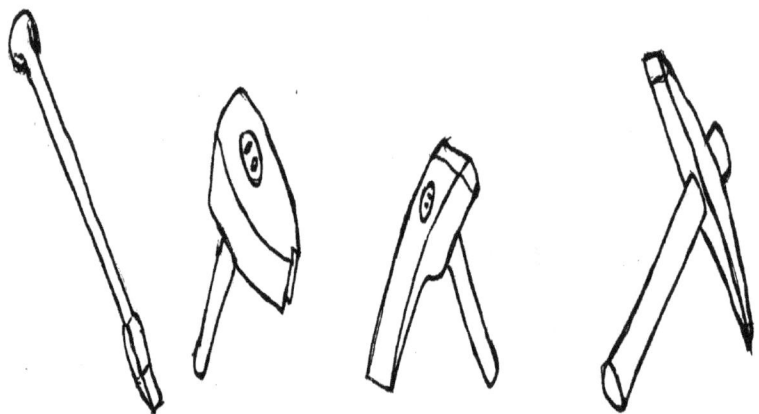

A long chisel bar (far left) for prying stones up can also help to shape them. So too can a stone cutter's pick (far right). The small mason's hammer (2ⁿᵈ from right) is used to chip off corners and protrusions. The heavy stone hammer (2ⁿᵈ from left) is used to fracture & break large stones. Chipping away or chiselling out a score line encourages the stone to break almost cleanly along that line. To a degree these can also be done by heating the stone and applying cold water. Another useful tool would be a sled or wheel barrow to pull the stone.

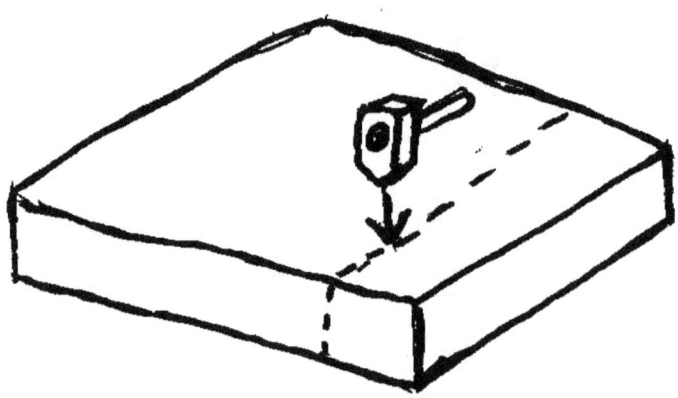

If you're making an insulated wall or one with open spaces for pipe or conduit, It's a good idea to build an outer and inner wall and link the two together with stone or wooden blocks. Alternately, you can use iron or steel rods that have been bent to better hold onto the stones and mortar.

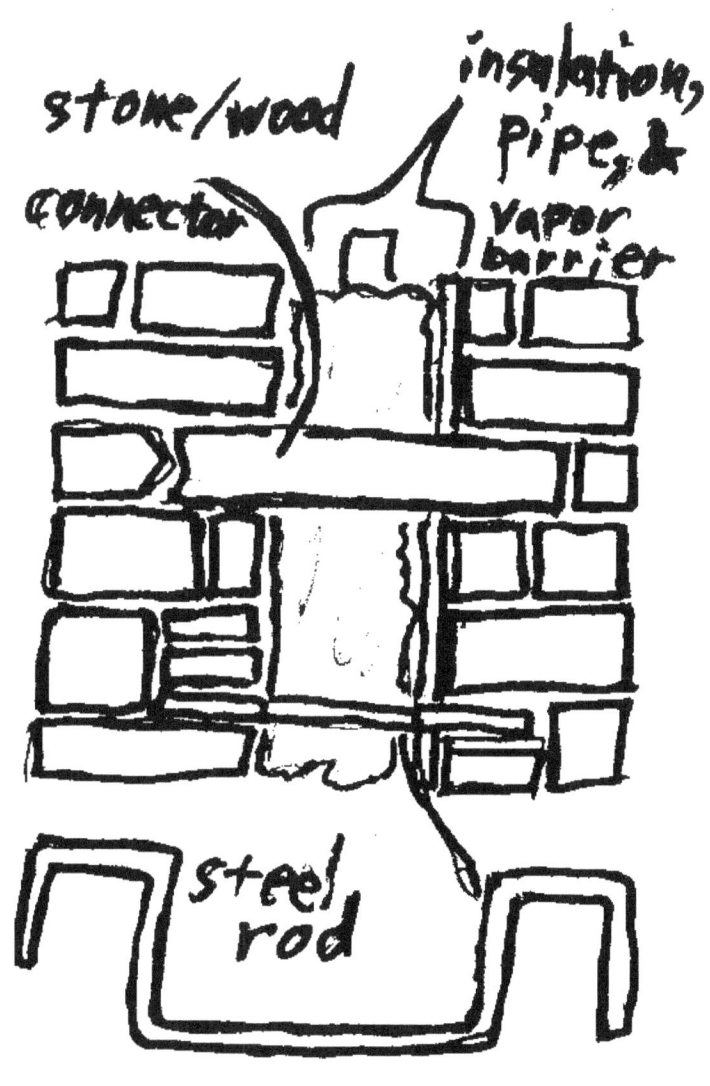

Metal Work:

Metal tools and building materials are becoming commonplace in many parts of the world. Their malleability (ability to be reshaped) and resilience in dealing with temperature extremes allows many option when used separately or in combination with other materials.

That being said, pure metal is hard to find in nature. It is usually chemically bonded with other elements in stone or salts that must be melted and sometimes processed in various ways before it can be used.

To use a common example, iron ore rocks are dug out of the ground either on the surface or from a mine. This is stacked in layers sandwiched between layers of charcoal or coke (heat refined coal) and heated past its melting point in a Cupola (from Latin & Greek meaning a cup) furnace. As it melts, oxygen and hydrogen from the ore bonds to the carbon and is released with the smoke. The liquid iron also picks up a lot of the carbon before it is cast into ingot moulds. The resemblance of the mould set-up to piglets suckling from a sow has given this Cast iron the common slang name of Pig iron.

Due to the high amount of carbon it has absorbed, solid cast iron is very hard and even brittle if it's thin or cold enough. This means it can't be hammered into shape. It has to be re-melted and re-cast into usable shapes.

When melted in a hot enough fire, especially if mixed with powdered lime or limestone, the carbon separates out leaving relatively pure iron known as Wrought iron. Solid wrought iron is very soft and easily bent and hammered into shape. It's also less prone to rusting. From there it can have some carbon re-added to make Carbon steel which can be hardened by heating and quickly cooling, softened (Annealed) by heating and slowly cooling, or Tempered for flexible tension by partially heating and cooling at a moderate rate.

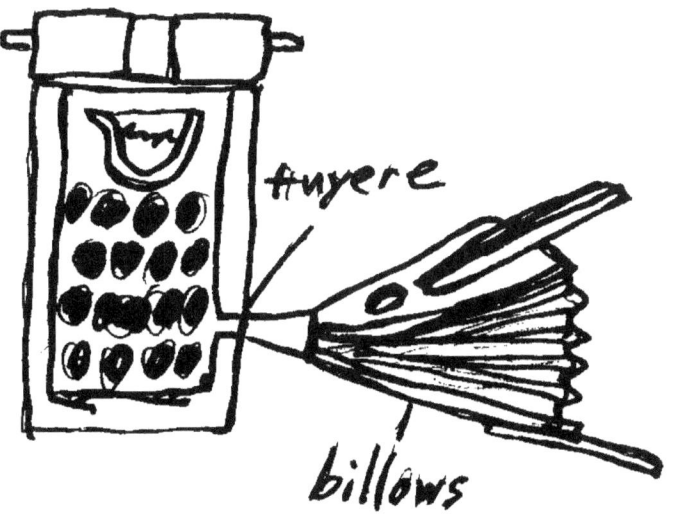

A forced air furnace with a crucible made in part from compressed fire clay is useful for melting and pouring metals. The hole for the air to move through is called a Tuyere.

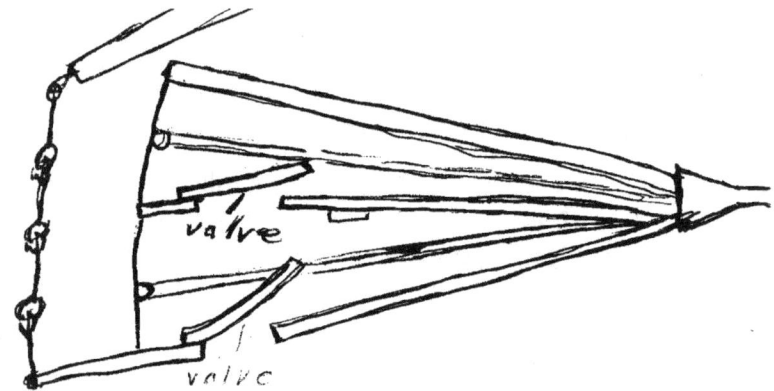

A Billows is one of the oldest methods of moving large quantities of air to increase the heat of a fire. There are many different designs with the common feature being a lining of flexible material and a flap that acts as a valve to control the direction of the airflow. Some complex, continuous flow designs like this "Great Billows", incorporated two valves.

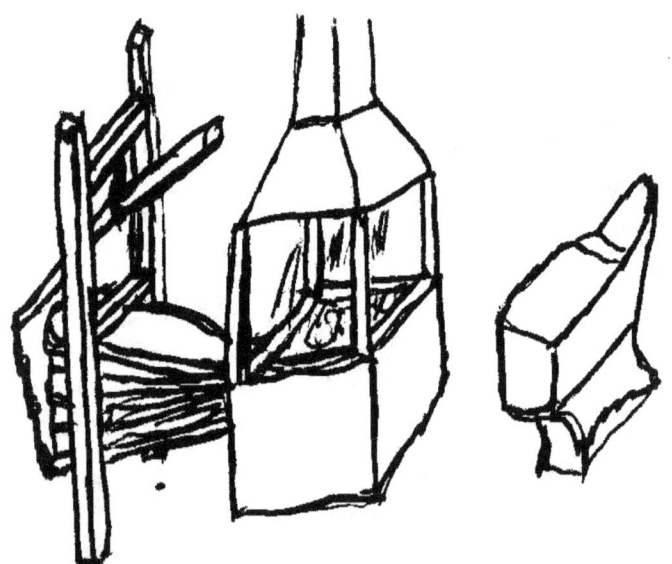

Without a closed cover on the furnace it can still be used in a smith's workshop heating metals to workable temperatures.

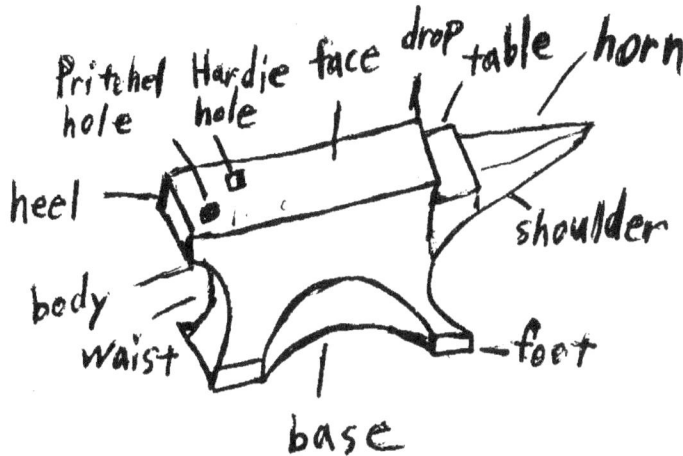

There have been many designs of anvils used by different cultures over the centuries. Even a smooth stone or slab of iron can do in a pinch, but different designs make it easier to turn out intricate works. One of the most popular has been the "Paris Anvil". All of the different surfaces provide different angles for bending and shaping metal.

hot Hardie   cold Hardie   bottom fuller

The Pritchel hole can be used to catch the end of a hot rod for bending it with a lot of leverage. The Hardie hole is used for holding a number of different anvil tools. The Hot Hardie and Cold Hardie are used to fold metals. The Bottom fuller is used to make a groove, increasing the overall size and strength of a piece without increasing its weight or mass.

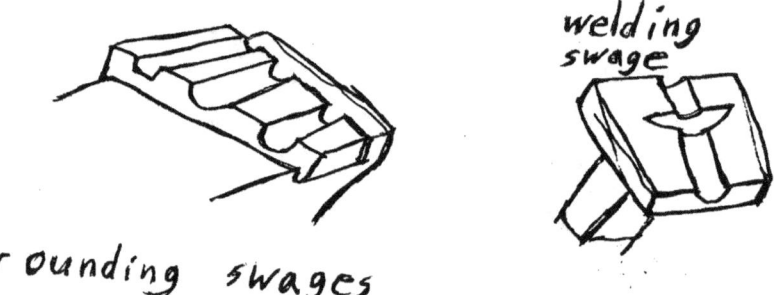

r ounding swages

Rounding swages are used to convert square rods into round rods or to turn sheet metal into rolled pipe which can then be welded or soldered. The Welding swage is used for joining two pieces of metal. A Monkey tool, Fork, or Hold fast tool can be used to hold an item in place while beating it.

The Anvil bick is used for making holes in pieces of metal and shaping rings. The ½ penny snub end scroll tool and the Rivet header are used for making specialized curves.

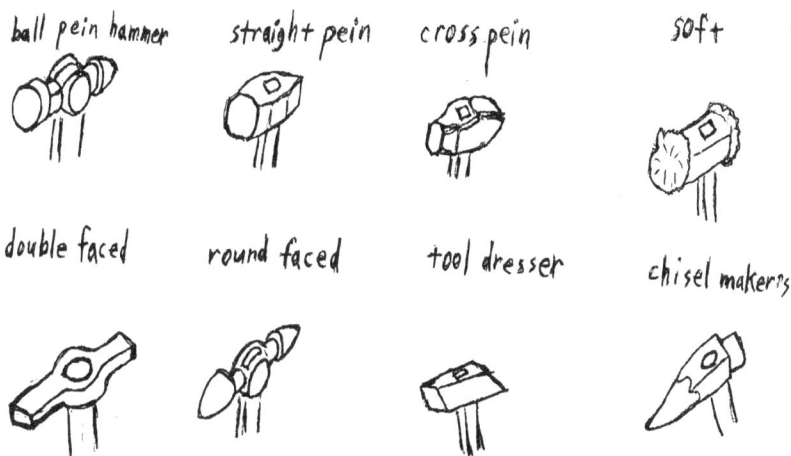

ball pein hammer   straight pein   cross pein   soft

double faced   round faced   tool dresser   chisel makers

There is a wide variety of hammers that are used for shaping metals. Each one has it's uses and advantages. Many smiths will use one or two types of hammers for all of their work. Others will make specialized hammers for each type of project they do. They can be softened to absorb shock and have just the striking surfaces hardened to hold them together.

There is also a variety of tongs and pliers for holding and manipulating hot objects. Again, many people will just use

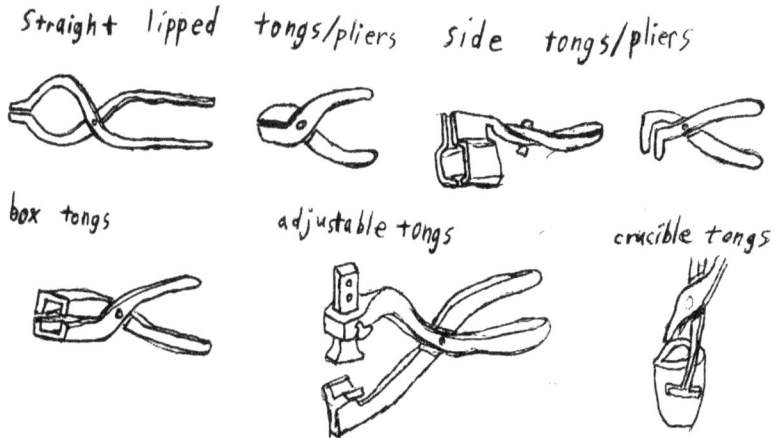

Straight lipped tongs/pliers   side tongs/pliers

box tongs   adjustable tongs   crucible tongs

one or a few types, but some will make specialized tools for every application.

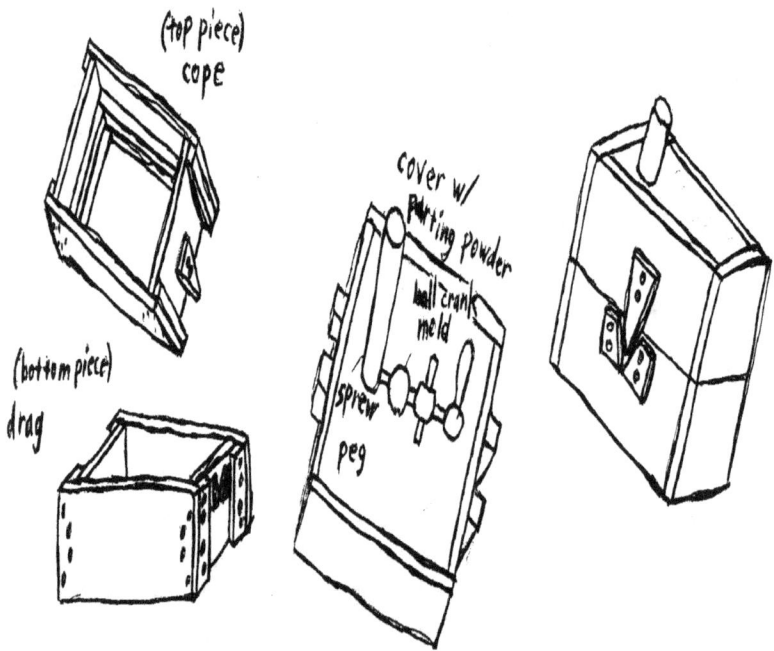

In order to make specialized shapes from molten metal, glass, or other similar substances, you need a mould for the shape to be cast in. This can be made from clay, sand, or plaster of Paris. If you want temporary moulds, try mixing clay and sand, 5-10% bentonite clay or 25-35% fire clay, and maybe 1-1 ½% flour and then enough water to keep its shape if squeezed.

Fill a bottom box (called a Drag) with that mix and press a shape that you want to make a copy of into the wet sand mix. Stick a thick dowel rod in the sand nearby and connect them with a thin dowel rod. Cover the model shape, connecting dowel, and all the sand with a parting powder. Powdered graphite, diatomaceous earth, or some other powder that won't gum up from the moisture will work. Place another box (called the Cope) to fit over the first. This should have ridges attached on the inside for the sand to stick to.

Pack the sand in tightly up to the top and push a stiff wire through to make a few vent holes. Remove the big dowel to make a hole to pour the molten material in (called a Sprue hole), and lift the top box gently and set it on it's side. If it falls apart, the sand may have been too wet or too dry, or you may not have been packed enough. If little pits of it fall through or pull out, there may not have been enough parting powder there.

Remove the model shape and the connecting dowels from the drag before replacing the cope. You might widen the top of the sprue hole to hold more molten material and act as a funnel. Let the material cool for a few hours, then use a stiff wire to check if the material is solid.

If need be, a metal object or block can be further shaped by clamping it in a vice and removing a chip at a time with a Metal chipping chisel.

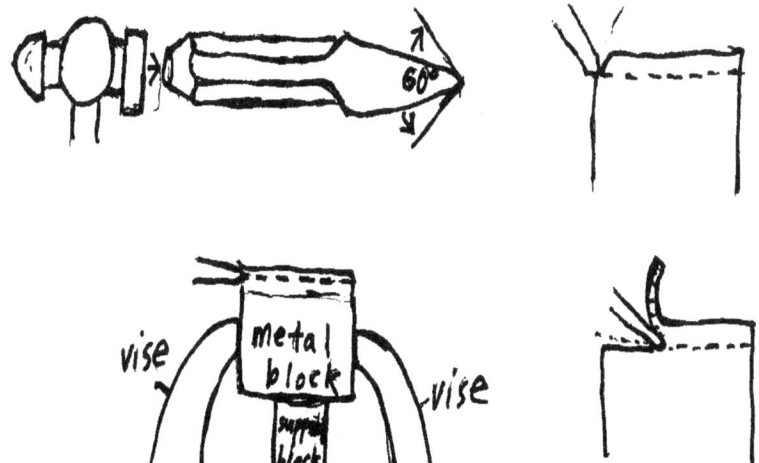

Remove large sections first, then you can do more intricate cuts.

Of course to make a mould you need an original shape to begin with. This shape can be made from wood, clay, wax, or any other material. In order to make precision shapes, a commonly used tool is the Machine lathe.

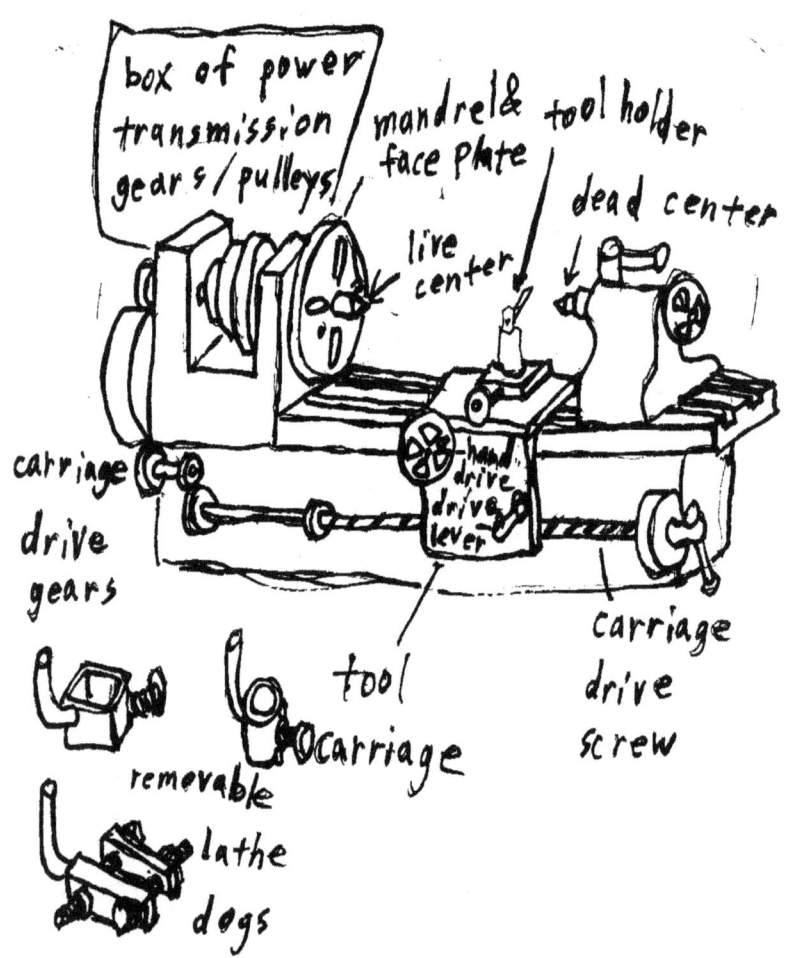

Gears or pulleys turn a faceplate at variable speeds. The faceplate has a mandrel system for holding a Live center point that supports and turns the material to be shaped. It can also support removable tools called Lathe dogs that help to support and secure the material to be shaped. A sliding base holds and positions the freely spinning Dead center point for holding the other end of the material to be shaped.

The tool holder and tool carriage can be positioned by hand turned knobs and levers, or the carriage can be geared in so that it move a certain distance to the side each time the faceplate/live center turns.

There are a variety of tools to be used with the machine lathe.

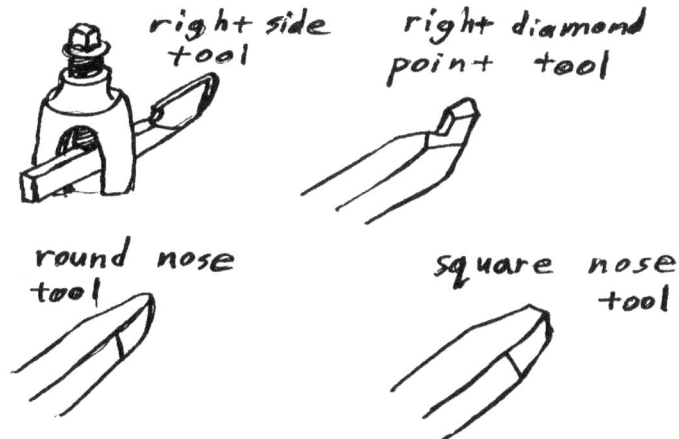

right side tool

right diamond point tool

round nose tool

square nose tool

Some of the most commonly used are the Right side tool for making deep cuts from right to left (there is a left side tool as well), the Right diamond tool for making shallow, precision cuts from right to left (and there is a left diamond tool too), the all purpose Round nose tool for shaping, and the Square nose tool for flattening. These tools can be thought of as wedges for separating the materials.

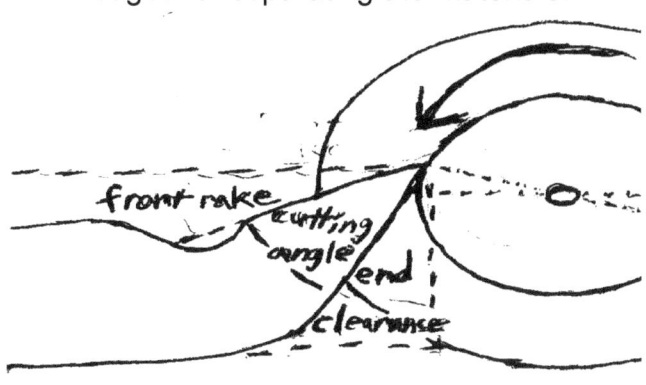

front rake

cutting angle

end clearance

The shape of the tool slopes back and down from the tip to form the "End clearance". If there isn't enough end clearance the tool won't cut very deeply. If there is too much, the tip may overheat, losing its sharpness. The shape also slopes back and down from the top to form the "Front rake". The steeper the front rake is the easier it is to quickly remove material. If it's too steep, again the tip may over heat or break. The angle of the tool's cutting surface is called the "Cutting angle" The best cutting angle depends on the material to be shaped. Steel and wrought iron can be cut with an angle between 40° and 50°. Tough cast iron on the other hand does better with an angle between 60° and 75°.

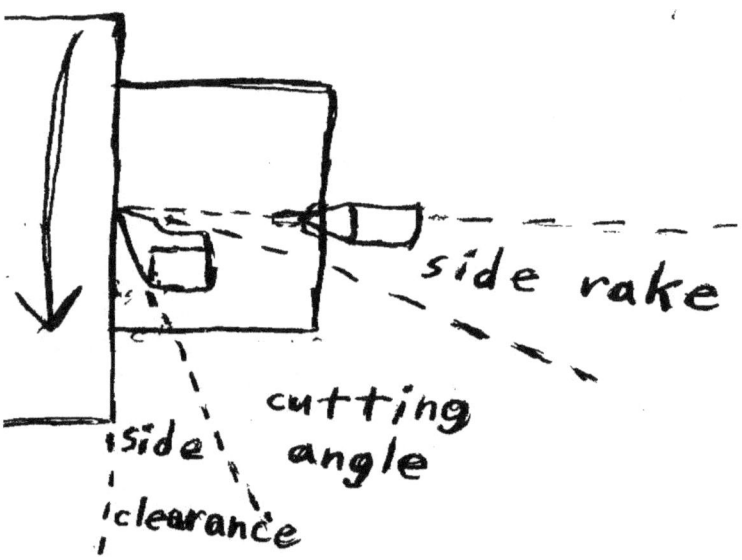

For cutting sideways, there is another cutting angle to give Side clearance and a Side rake. Again there are the same angles and considerations as for the front cutting angle.

There are also specialized tools for cutting screw threads. The most common internationally is the 60° tool for cutting Sharp "V" shaped threads.

60° Sharp V threading tool

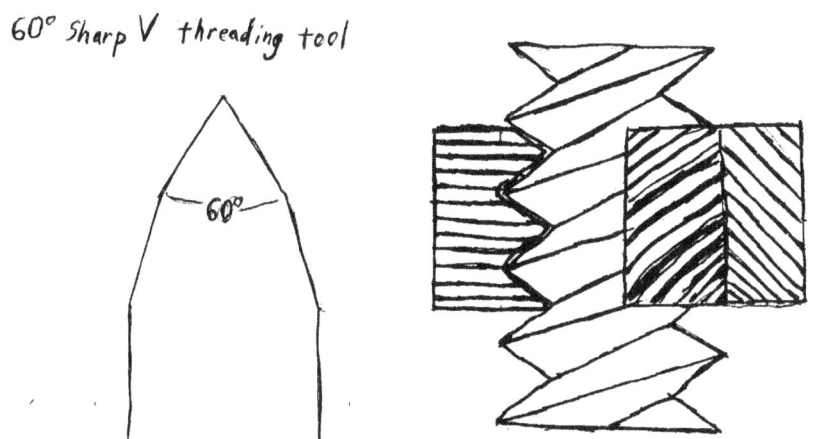

For the threads to line up properly the tool carriage needs to move to the side exactly the width of the thread for each time the live center turns.

The most common for the United States is the U.S. Standard or USS threading tool. It still has the 60° angle, but the point has been ground off. For these threads to line up properly the carriage needs to move a   distance equal to the width of the thread plus the width of the ground off tip. The main benefit of this type of thread is the reduced chance of cutting one's self on the threads.

U.S. Standard threading tool

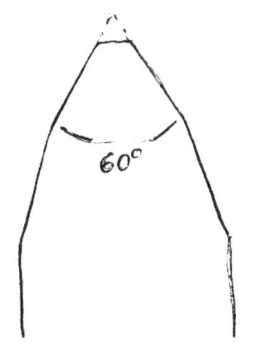

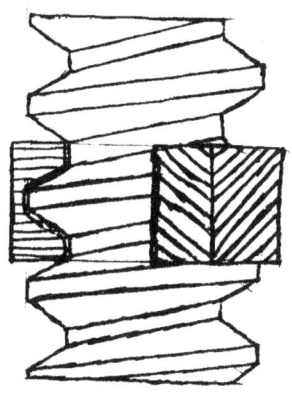

For cutting threads inside of a nut or sleeve or shaping the inside of a pipe or other cylinder there are boring tools. These need a greater end clearance for working in such a narrow space.

Aside from those two standards, there are a number of other threading patterns that are used in various parts of the world and in various industries. One of those is the "Square thread". The tool carriage must move twice the thread width for every turn of the live center for this one. The main benefits are thread strength and reduced risk of cross threading.

Lathe tools can be forged on an anvil out of carbon steel and tempered and hardened to the point where they are "Glass hard", meaning to the point where they can't be scratched with a filing tool. It may be a good idea to anneal the materials to be shaped on the lathe and re-temper/ harden them later.

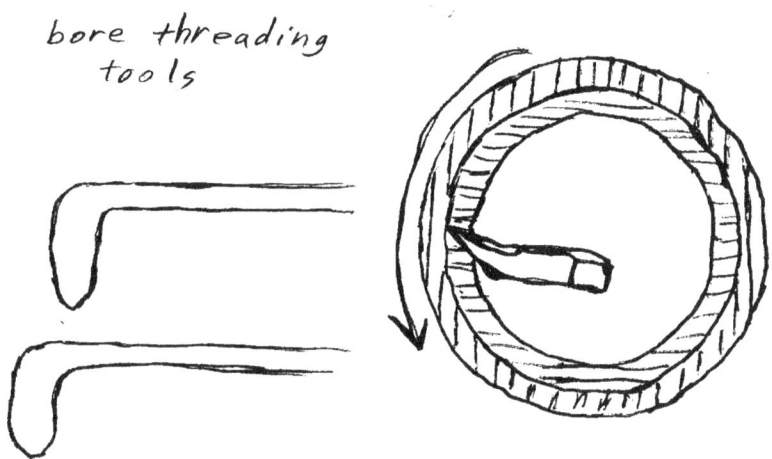

bore threading tools

Some lathe tools are made out of high-speed steel. "High-speed steel" is an iron alloy with 0.25% vanadium and either: tungsten, molybdenum and tungsten, or chromium and molybdenum. Those tools can keep their cutting edge until the tool is red hot, allowing more work to be done faster. High-speed steel is hardened by raising the temperature to white heat (~2100°F/~1150°C) and cooling them with a blast of cool air or in oil, never in water. To anneal it, pack it in sand in a cast-iron box, heat it to white-hot, and let it cool slowly. To temper it, dip it in oil heated to 460°F/238°C and cool it in a blast of forced air. In order to forge tools out of high-speed steel it must be heated to a bright lemon yellow color.

Of course machine lathes aren't usually very portable, so there are Die stocks and Boring tools with special attachable wrenches that can be used to thread pieces of pipe to be connected together. Some of them are even adjustable for different pipe sizes.

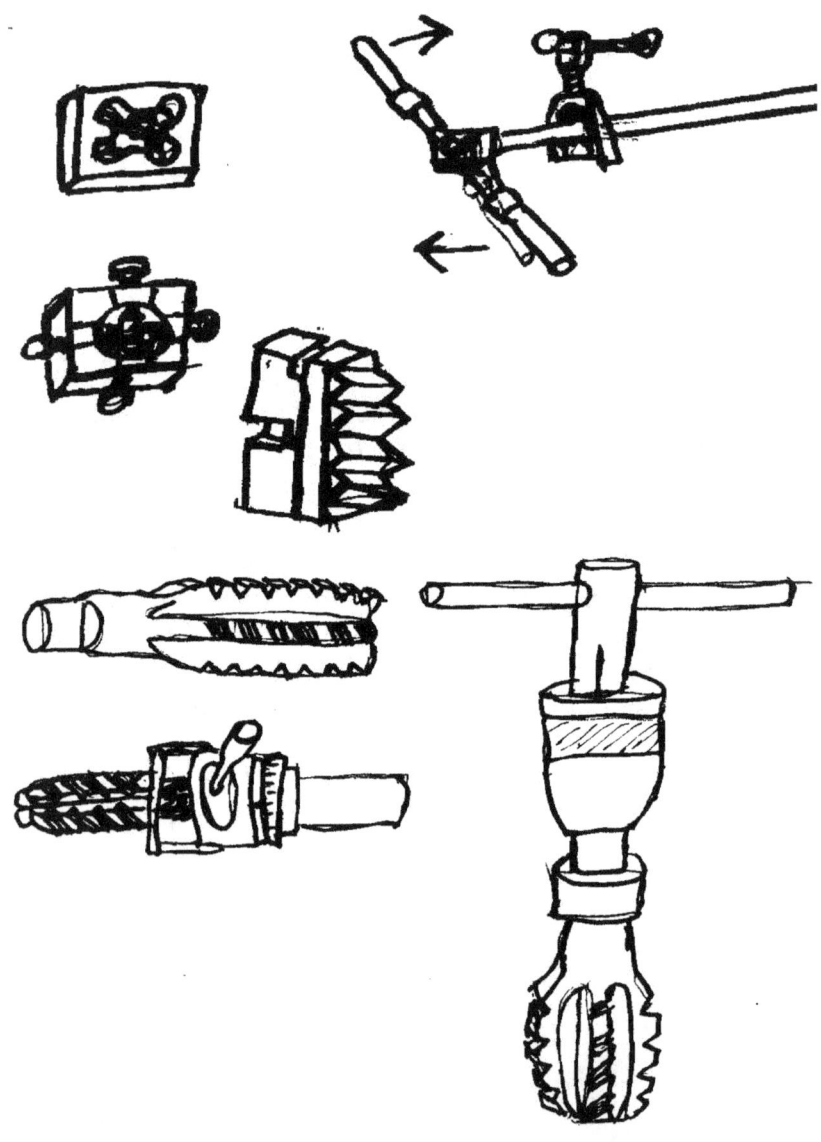

Other than threading a screw/bolt and nut there are a number of, more permanent, ways to join pieces of metal.

Using a flammable compressed gas or volatile flammable liquid   especially along with compressed oxygen/air to make it burn hotter, is a fairly old way of applying heat to weld, cut, solder, or braze metals. This is often called an Oxyacetylene system, but other flammable gasses can be used. Caution: Compressed oxygen can cause oil; petroleum; grease; & other fats to explode. Be sure to clean your work.

It's important to balance out the fuel gas with the oxygen for the hottest stable flame. The basic gas flame has a double cone coming out of the aperture and a long feathering flame that may produce significant smoke. Increasing oxygen shortens the feathers and the outer cone until it merges with the inner cone. Too much oxygen/insufficient gas produces a tiny cone and a squealing noise.

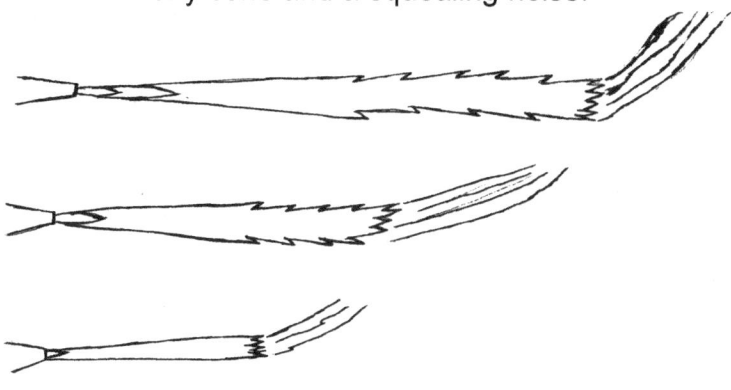

Apply the flame to a sheet of metal to the point where the tip of the cone comes close to the metal and move it in small circles until a small puddle of molten metal forms. You can dip the tip of a filler/welding rod into this puddle and melt the tip off into the puddle. Keep up the circular action as you move across the sheet of metal, preheating the next section as you melt that edge of the puddle and let the other edge cool and solidify. This leaves a line of overlapping weld beads

Set two sheets of metal edge to edge on heat resistant supports with very little space between them. Since they will expand to touch each other and thicken, it's a good idea if they are tapered where they meet.

If the weld bead penetrates to half or more of the thickness of the metal plates, it's considered to be sufficient. If it drops through past the other side, it's still considered acceptable.

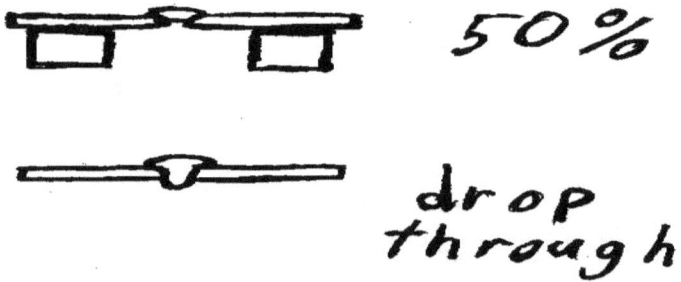

You can also prop a couple of sheets against each other at an angle, tack them together by making a couple of weld beads near the ends and in between to keep the heat from warping them, and then weld them from one end to the other. You can even do this without a filler rod by just working the puddle from one side to the other. It's convenient to think of the metal as ice that melts and refreezes.

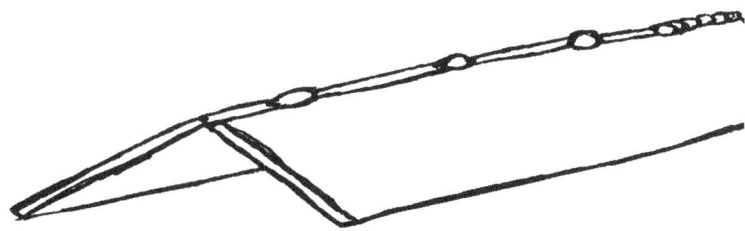

In place of a normal welding torch you can use a cutting torch. In addition to controls for fuel gas and oxygen there's a trigger pressure control for unleashing a burst of high pressured oxygen. Properly balanced and using the right technique and the right tip for the metal thickness this will burn the metal in a line of fine cracks. Even without having everything perfect, you can melt a section and blow the liquid metal out of it. The cutting tips have smaller holes around a center aperture. These form a hub of mini flame cones that preheat the metal. Angling the torch so that one of these is pointed in the direction you want to cut allows the job to be done faster and more efficiently.

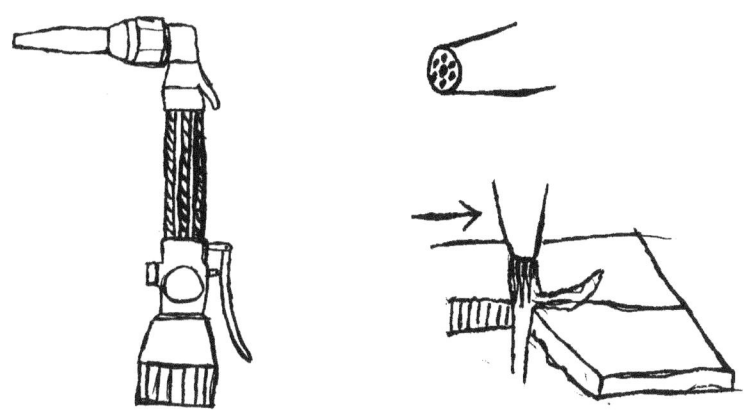

Metal Thickness	Welding Tip Size #	Cutting Tip Size #
1/64"-1/32" (.39mm-.78mm)	000/AW200/.020	
1/32"-3/64"(.78mm-1.17mm)	00/AW201/.025	
1/32"-5/64"(.78mm-1.95mm)	0/AW203/.035	
3/64"-3/32"(1.17mm-2.34mm)	1/AW204/.040	
1/16"-1/8"(1.56mm-3.13mm)	2/AW205/.046	OO
1/8"-3/16"(3.13mm-4.69mm)	3/AW207/.060	O
3/16"-1/4"(4.69mm-6.25mm)	4/AW209/.073	O
1/4"-1/2"(6.25mm-1.25cm)	5/AW210/.090	1
1/2"-3/4"(1.25cm-1.88cm)		1
1"-2"(2.5cm-5cm)		2
4"-5"(10cm-12.5cm)		3
6"(15cm)		4

Metal thinner than 1/16 inch should be cut with other techniques. If it's thicker than 1/2 inch, gas welding isn't very efficient. You'll need a hammer and a wire brush to clean up the slag, flux, and any other contaminants from the welded joint.

There are also "Rosebud" tips for softening and bending metal.

Brazing and soldering are metal joining methods that rely on the surface tension of a filler material rather than melting the base metal.

Soldering is done below 800°F(427°C) and usually uses an alloy of lead and tin or silver and tin along with a rosin or acid based flux (may be in liquid or paste form) to de-oxidise and clean the base metal to allow the filler to stick to it.

Brazing is done below the melting point of the base metal, usually below 1500°F(816°C), using filler rods that are usually made of brass or bronze. These may be preheated and dipped in a powdered flux (anhydrous copper sulphate is often used) prior to use, or they may come pre-coated with flux.

Metal sheets to be brazed should be overlapped instead of butting them together. This allows for increased surface tension that prevents the joint from breaking. If you must have a smooth surface you can overlap the butted joint with another strap of metal. Pre-sanding and roughing up the surface maximizes the strength of the joint. These processes are often used to join pipes at different angles.

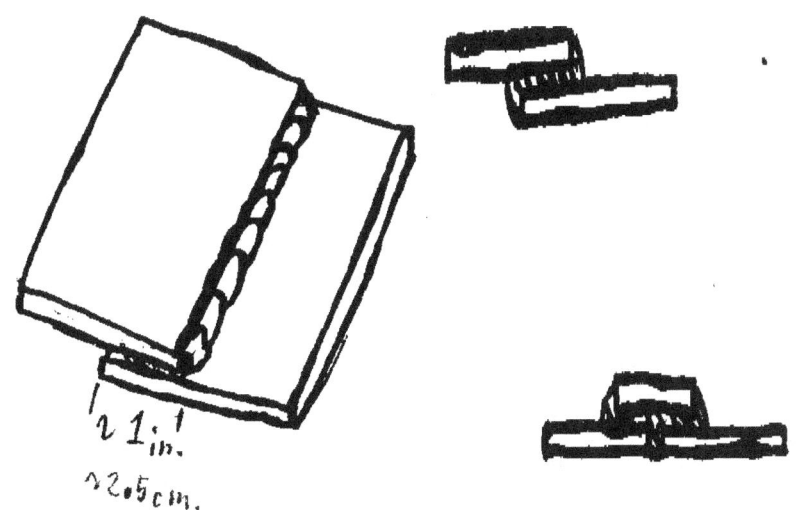

There are a number of specialized tools that have been made for soldering called Soldering irons. One of the oldest was a pointed block of bronze/copper covered metal on a wood handled shaft. This could be heated in a flame and applied to the base metal around a joint to raise the temperature enough to melt the solder filler material (solder for short). The hotter it gets the better the solder flows into and around the joint. One of my old teachers used to say: "The solder flows where the heat goes."

For smaller pieces of metal a convenient pistol shaped, electrically heated, "Soldering-gun" was developed. For delicate work like jewellery and electronic circuits, a pencil shaped electric soldering iron was developed. Sometimes applying some liquid flux from a squeeze bottle to the joint beforehand helps to get the solder to flow in as the liquid evaporates. That same teacher used to say: "Solder flows where the flux goes."

A number of methods have been developed for welding using electricity. The most basic of these would be Arc welding. A high amperage power supply is grounded to the base metal and an electrode holder, cable   connected to the output of that power supply, is clamped to a flux covered welding rod (also called a stick electrode).

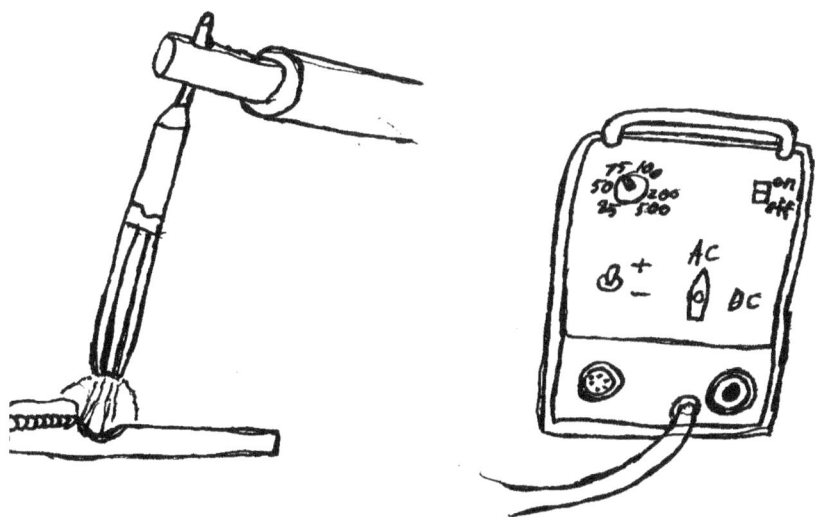

The tip of this welding rod is brought close enough to the base metal to start an arc(like a lightning bolt) of electricity. It may be necessary for that tip to be touched, tapped, or dragged across the base metal to get it started. This is sometimes referred to as striking it. The arc heats the base metal to form a puddle and shoots off bits of the electrode as sparks. At the same time the heat vaporizes the flux near the end to form a gas shield that drives oxygen, moisture and other contaminants away from the puddle.

Different electrodes may be made of different material and be coated with different flux materials. Each type may work better at different angles and using current of different or alternating polarities.{The principles of   electricity will be discussed in more detail in another book.}

All of these electrical welding methods produce intense ultraviolet light/radiation, the same light/radiation from the sun that can give you a sunburn or damage your eyes. As such it's important to protect your eyes and skin by using a welding helmet with the appropriate shade of lenses and appropriate clothing including leather gloves and shoes with rubber soles.

The next on the list should probably be Wire-feed welding, also known as MIG which stands for metal-inert-gas or metal-in-gun depending on the manufacturer. An electrically charged wire filler electrode and an inert gas (such as helium or any of the other "noble gases") are fed through a hose to a "gun" applicator when the trigger switch is squeezed. The wire may or may not be hollow with flux inside it. The end of the wire melts as the arc forms the puddle and the gas pushes out contaminants. This tends to be messier than regular arc welding, but it goes by faster, allowing projects to be finished much quicker.

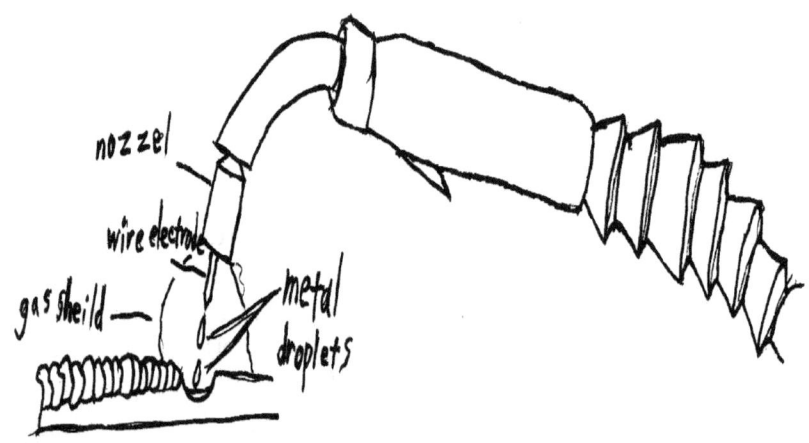

Next up would be TIG Welding which stands for tungsten-inert-gas or tungsten-in-gun. It has also been called Heli-Arc after the trademark name of one manufacturer. A tungsten or tungsten-alloy electrode is fitted in a holder called a torch that is connected by a hose/cable to a power supply and usually an inert gas supply. The basic power supply starts the arc with a high voltage and high frequency current that repels oxygen. It can be set to maintain the high frequency or, as long as you have the inert gas flowing, switch to a lower frequency or direct current which heats the base metal more efficiently.

The TIG arc is usually bigger than that of MIG or regular arc welding,   allowing you to dip a filler rod in it as though it was a torch flame. There are no sparks or splatters flying about with TIG welding, and almost no slag. On the other hand, the metal needs to be pre-cleaned or you'll lose you arc. It may also be necessary to use a copper striking plate to get the arc started.

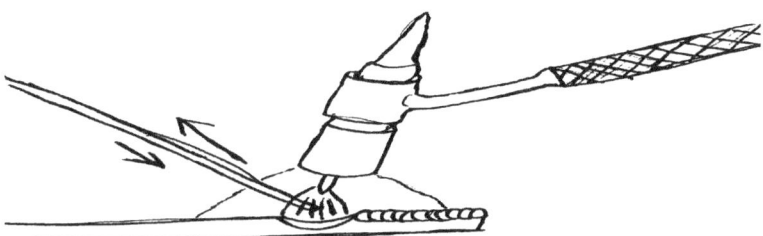

Sharper tips that look like pencils produce hotter arcs and are used for iron and steel. Needle shaped points are for fine/detailed work and crayon shaped points are used for aluminum. The tips often melt and ball up after doing a certain amount of work and you'll need to re-grind them into the appropriate shape. Always grind it lengthwise or it will have an arc pattern that is difficult to control.

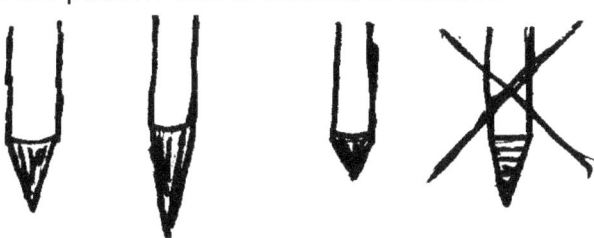

There is also a way of cutting with electricity called Plasma-Arc Cutting. An electrode is located within a torch nozzle with a orifice that constricts the arc while an inert gas (usually argon); nitrogen; or just dry air is forced through it and energized into a plasma state with temperatures up to 50,000°F(27,760°C). This instantly melts the metal and blows it out in very narrow lines, allowing the metal to be cut very quickly with no pre-cleaning. If plan to use TIG welding afterwards, you will need to clean the metal with acid if you use nitrogen or dry air. {We will discuss the plasma state of matter in detail in the next chapter.}

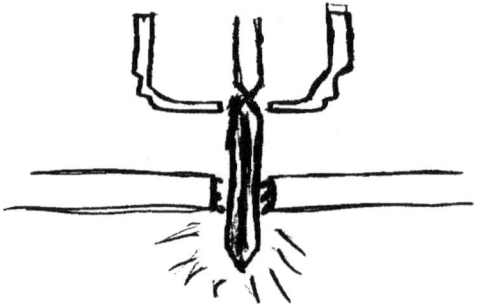

It is also possible to spot-weld even dissimilar metals by clamping them together and running an arc through the point where they touch.

7 - Afterword:

I hope you enjoy the information and skills this volume offers.

Join us in the next volume for lessons on matter, ballistics, and fluid dynamics.

8 - References

The following references were helpful in writing this book:

"Welder's Handbook * A Guide to Plasma Cutting, Oxyacetylene, Arc, MIG, and TIG Welding" by Richard Finch

"Thinking Physics * Practical Lessons In Critical Thinking" by Lewis Carol Epstein

"Schaum's Outline Series * Theory And Problems Of College Physics 8/ed" by Frederick J. Bueche

"Experimental Science * Elementary Practical and Experimental Physics Volume One & Two" by George M. Hopkins

"Mr. Wizard's Supermarket Science * More than 100 fascinating and fun experiments using easy-to-find everyday items" by Don Herbert

"Building With Stone" by Charles McRaven

"Shelters, Shacks, And Shanties * The Classic Guide to Building Wilderness Shelter" by D. C. Beard

"Elements Of Machine Work" & "Advanced Machine Work" by R. H, Smith

"Build Your Own Metal Working Shop From Scrap" series by David J. Gingery

"Math Magic * The Human Calculator Shows How To Master Everyday Math Problems In Seconds" by Scott Flansburg

"The McCall's Book of Handcrafts" by Random House

"Practical Mathematics * for Home Study Arithmetic, Geometry, Algebra, & Trigonometry" by C. I. Palmer

"Survival Wisdom & Know-How * Everything You Need to Know to Subsist in the Wilderness" by Stackpole Books

"Simple Scientific Experiments * How to Perform Entertaining and Instructive Experiments with Simple Home-made Apparatus" by Aurel de Ratti

"The SCIENTIFIC AMERICAN Book of Projects for THE AMATEUR   SCIENTIST" by C. L. Stong

"The Marriam-Webster Dictionary" by Marriam Webster

Alphabetical Index